Instructor's Manual

to accompany

NUMERICAL METHODS

By J. Douglas Faires and
Richard L. Burden

PREPARED BY

J. DOUGLAS FAIRES AND
RICHARD L. BURDEN

Youngstown State University

PWS-KENT Publishing Company Boston

PWS-KENT
Publishing Company

20 Park Plaza
Boston, Massachusetts 02116

The computer disk that accompanies this Instructor's Manual is intended for users of Numerical Methods by Faires and Burden. It may not be copied, sold, or redistributed without the permission of PWS-Kent Publishing Company.

Reviewers of Numerical Methods by Faires and Burden have permission to make use of this package for a period of 90 days beginning from the day the package is received.

Printed in the United States of America

1 2 3 4 5 6 7 8 9 -- 97 96 95 94 93

PREFACE

This Instructor's Manual for *Numerical Methods* by Faires and Burden contains solutions to all the exercises in the text. Although the answers to the odd exercises are also in the back of the text, the results listed in this Instructor's Manual often go beyond those in the book. We do not, for example, place in the book the long solutions to theoretical and applied exercises. You will find them here.

In addition to the answers to the exercises you will find here a listing of the instructions for the program disk that accompanies *Numerical Methods* together with a copy of the disk that contains the programs, both in Pascal and in FORTRAN.

We hope you find these items useful and that our supplement package provides flexibility for instructors teaching from *Numerical Methods.* If you have any suggestions for improvements that can be incorporated into future editions of the book or the supplements, we would be most grateful to receive your comments.

As a final remark to this preface, we would like to express our deep appreciation to Sharyn Campbell for her help in preparing this material. She has been one of the best of a long line of outstanding student assistants that we have had the pleasure of working with at our institution.

Youngstown State University — J. Douglas Faires

November 13, 1992 — Richard L. Burden

Table of Contents

Numerical Methods Programs

Solutions to the Exercises

NUMERICAL METHODS PROGRAMS

I. About the Program Disk

The disk in this package contains a Pascal and a FORTRAN program for each of the methods presented in the book *Numerical Methods* by Faires and Burden. Each program is illustrated with a sample problem or example closely correlated to the text. This permits the program to be run initially to see the form of the input and output and then be modified for other problems by making minor changes in the program.

The programs are designed to run on a minimally configured computer using the DOS platform. To use the program disk you must have the PC DOS or MS DOS operating system and an appropriate Pascal or FORTRAN compiler.

II. Running the Pascal programs

The following instructions for are appropriate for running the Pascal programs if you are using Version 4.0 or higher of the Borland Turbo Pascal Compiler. If you are using some other version of Pascal you should consult your Pascal compiler manual for the equivalent Pascal commands.

1. Boot the system.
2. Place the Turbo Pascal disk in drive `A` and make `A` the current drive.
3. Invoke Turbo Pascal by entering the command

   `TURBO`

4. You will see a line on the screen that reads

   `FILE   EDIT   RUN   COMPILE   OPTIONS`

5. The presence of a numeric coprocessor can be indicated by using `OPTIONS`.
6. Enter the command

   `F`

   for File.
7. You will then see a submenu listing the options

```
LOAD
PICK
NEW
SAVE
WRITE TO
DIRECTORY
CHANGE DIR
OS SHELL
QUIT
```

8. If you have two disk drives, place the Numerical Methods disk in drive **B** and change the logged drive to **B**. This is accomplished by selecting the option

   CHANGE DIR

   and entering

   B:

   If you have only one disk drive, remove the Turbo Pascal Compilier disk from the drive and replace it with the appropriate Numerical Methods disk.
9. Enter the command

   L

   for Load and the screen will show

   LOAD FILE NAME
   ***.PAS**
10. Type the command to execute the appropriate program. For example, if you want to run the program for the Bisection method type

    BISECT21

    over the line ***.PAS** and the first 14 lines of the file **BISECT21.PAS** will appear on the screen.
11. After the program has been loaded into the computer, enter the command

    ^KD
12. Enter the command

    R

    to run the program.
13. To exit from Turbo Pascal, depress the **x** or **X** key while simultaneously depressing the **Alt** key or choose **QUIT** under the **FILE** option.

The following steps can be followed to modify one of the Pascal programs:

1. From within Turbo Pascal and after selecting the program as described above, select the option **E** for edit.
2. The program will appear on the screen for editing.
3. The following editing instructions and commands can be used to modify and save the new program. The Turbo Pascal editor is a simplified version of the Wordstar editor. If you are familiar with that common word-processing program, you will find that your Wordstar commands are generally appropriate.
   (a) To exit the editor enter the command **^KD**
   (b) Use the cursor keys (←, →, ↑, and ↓) to move around within the file.
   (c) Letters typed will be inserted and not overwritten unless the **Ins** key is depressed. This key is used to toggle between the insert mode and the overwrite mode.
   (d) The command **^T** deletes the next word to the right and the command **^Y** deletes the current line. The command **^QY** deletes the rest of the line to the right of the cursor. These are only a few of the most frequently used editing commands. Please consult your Turbo Pascal manual if you need further information.

(e) Depressing the **ENTER** or **RETURN** key (often denoted on the keyboard by ↩) at the end of a line will insert a blank line following the current line.

(f) Comments are inserted in the program to indicate lines that must be changed in order to change the problem. (Note: Comments are enclosed within { }.)

4. After leaving the editor, the program as modified can be run using the command **R**.
5. The modified program can be saved on the Numerical Methods disk in place of the previous version.
6. Turbo Pascal saves a back-up copy of a program that is the working file in case the editing session is a failure. The back-up copy has the same name as the working file except that the extension **.PAS** is changed to **.BAK**.

III. Running the FORTRAN programs

The FORTRAN programs are designed to run correctly under any FORTRAN compiler for FORTRAN77. The commands given below are specific to Microsoft FORTRAN Version 5.1 for the MS-DOS or PC-DOS operating systems. Modifications to the commands may be required if another compiler or operating system is being used.

Suppose that we want to run the FORTRAN program for the Bisection method, BISECT21.FOR. To compile the ASCII file for BISECT21.FOR, enter the command

```
FL /Fs  BISECT21.FOR
```

This command combines compilation and linking and produces the file BISECT21.LST for human consideration and, if no errors are found, the file BISECT21.EXE for computer execution. Any errors found during compilation appear on the terminal screen as well as in the file BISECT21.LST. Once successful compilation and linking has occurred, the program can be run by issuing the command

```
BISECT21
```

To alter a program file you can use either the integrated environment provided with Microsoft FORTRAN 5.1 or an editor or word processor that creates an ASCII character file.

IV. Description of the Programs

The following pages briefly describe the methods on which the programs in this package are based. Each method is listed together with the name of the associated program and the page in Numerical Methods on which the algorithm for the method can be found. A description of the sample problem for the program is provided together with typical values for the input that is requested within the program. The programs are stated using the generic extension **.EXT** , which should be replaced by **.PAS** , if you are using a Pascal compiler, and by **.FOR** , if you are using FORTRAN.

During the execution of some of the programs you will be asked questions of the form

```
Has the function F been created in the program immediately
        proceeding the INPUT procedure ?
```

To run the sample problems you should enter the response

```
Y (for Yes)
```

for the Pascal programs or

```
'Y' (for Yes)
```

for the FORTRAN programs in each instance, since the functions are embedded within the programs. The functions will need to be changed, however, if the programs are modified to solve other problems.

The output data can be extensive from some of the programs. When this is likely to be the case, the programs have been constructed so that it is convenient to place the output directly into an output file. The screen will display

```
Choice of output method:
1. Output to screen
2. Output to text file
Please enter 1 or 2
```

To send the output to a data file type the response

```
2
```

The screen will now request the name of the output file to which the data is to be sent

```
Input the file name in the form - drive:name.ext
```

to which the response should be a filename such as

```
FILENAME.OUT
```

for the Pascal programs or

```
'FILENAME.OUT'
```

for the FORTRAN programs. The file `FILENAME.OUT` can then be examined to determine the output from the program.

Some of the programs require the input of large amounts of data. To enable the programs to be run quickly and efficiently, the data can first be placed in data files and the data files read by the program. For example, when running the program for Neville's method, `NEVLLE31.EXT` , using the defined data file `NEVLLE31.DTA` for the sample problem, you will first see a screen which states:

```
Choice of input method:
1. Input entry by entry from the keyboard
2. Input data from a text file
3. Generate data using a function F
Choose 1, 2, or 3 please
```

If you choose `1` you will need to enter all the data for the program into the keyboard. Choosing `3` will require that a function be defined that will be used to provide the data. With this choice the program will terminate until the program is modified to incorporate the function.

To run any of the sample problems when this choice occurs you should enter

`2`

The screen will now display the question

```
Has a text file been created with the data in two columns ?
Enter Y or N
```

Answer this question with the response

`Y (for Yes)`

for the Pascal programs or

`'Y' (for Yes)`

for the FORTRAN programs. The screen will now display a request for the name of the input data file:

```
Input the file name in the form - drive:name.ext
for example:   A:NAME.DTA
```

The correct response to this request for the sample problem using Neville's method is

`NEVLLE31.DTA`

for the Pascal programs or

`'NEVLLE31.DTA'`

for the FORTRAN programs.

The program will now request the remaining data from the keyboard. In the case of Neville's method it will request the degree `N` of the interpolatory polynomial and the number `x` at which the polynomial is to be evaluated. These values are shown in the input for the program.

PROGRAMS FOR CHAPTER 2

BISECTION METHOD "BISECT21.EXT" PAGE 27

This program uses the Bisection Method to approximate a root of the equation $f(x) = 0$ lying in the interval $[a, b]$. The sample problem uses

$$f(x) = x^3 + 4x^2 - 10.$$

INPUT: $a = 1, \quad b = 2, \quad TOL = 5 \times 10^{-4}, \quad N_0 = 20$

SECANT METHOD "SECANT22.EXT" PAGE 32

This program uses the Secant Method to approximate a root of the equation $f(x) = 0$. The sample problem uses

$$f(x) = \cos x - x.$$

INPUT: $p_0 = \frac{1}{2}, \quad p_1 = \frac{\pi}{4}, \quad TOL = 5 \times 10^{-4}, \quad N_0 = 15$

METHOD OF FALSE POSITION "FALPOS23.EXT" PAGE 34

This program uses the Method of False Position to approximate a root of the equation $f(x) = 0$. The sample problem uses

$$f(x) = \cos x - x.$$

INPUT: $p_0 = \frac{1}{2}$, $p_1 = \frac{\pi}{4}$, $TOL = 5 \times 10^{-4}$, $N_0 = 15$

NEWTON'S METHOD "NEWTON24.EXT" PAGE 37

This program uses Newton's Method to approximate a root of the equation $f(x) = 0$. The sample problem uses

$$f(x) = \cos x - x \quad \text{with} \quad f'(x) = -\sin x - 1.$$

INPUT: $p_0 = \frac{\pi}{4}$, $TOL = 5 \times 10^{-4}$, $N_0 = 15$

MÜLLER'S METHOD "MULLER25.EXT" PAGE 46

This program uses Müller's Method to approximate a root of an arbitrary polynomial of the form

$$f(x) = a_n x^n + a_{n-1} x^{n-1} + \ldots + a_1 x + a_0.$$

The sample problem uses

$$f(x) = 16x^4 - 40x^3 + 5x^2 + 20x + 6.$$

INPUT: $n = 4$, $a_0 = 6$, $a_1 = 20$, $a_2 = 5$, $a_3 = -40$, $a_4 = 16$,
$TOL = 0.00001$, $N_0 = 30$, $x_0 = \frac{1}{2}$, $x_1 = -\frac{1}{2}$, $x_2 = 0$

PROGRAMS FOR CHAPTER 3

NEVILLE ITERATED INTERPOLATION "NEVLLE31.EXT" PAGE 59

This program uses Neville's Iterated Interpolation Method to evaluate the nth degree interpolating polynomial $P(x)$ on the $n+1$ distinct numbers $x_0, \ldots, x_n$ at the number x for a given function f. The sample problem considers the Bessel function of the first kind of order zero at $x = 1.5$.

INPUT: NEVLLE31.DTA, $n = 4$, $x = 1.5$

NEWTON INTERPOLATORY DIVIDED-DIFFERENCE FORMULA "DIVDIF32.EXT" PAGE 63

This program uses Newton's Interpolatory Divided-Difference Formula to evaluate the divided-difference coefficients of the nth degree interpolatory polynomial $P(x)$ on the $n+1$ distinct numbers $x_0, ..., x_n$ for a given function f. The sample problem considers the Bessel function of the first kind of order zero.

INPUT: DIVDIF32.DTA, $n = 4$

HERMITE INTERPOLATION "HERMIT33.EXT" PAGE 72

This program uses Hermite's Interpolation Method to obtain the coefficients of the Hermite interpolating polynomial $H(x)$ on the $n+1$ distinct numbers $x_0, ..., x_n$ for a given function f. The sample problem considers the Bessel function of the first kind of order zero.

INPUT: HERMIT33.DTA, $n = 2$

NATURAL CUBIC SPLINE INTERPOLATION "NCUBSP34.EXT" PAGE 78

This program uses the Natural Cubic Spline Method to construct the free cubic spline interpolant S for a function f. The sample problem considers $f(x) = e^{2x}$ on the interval $[0, 1]$.

INPUT: (Select input option 2.) NCUBSP34.DTA, $n = 4$

CLAMPED CUBIC SPLINE INTERPOLATION "CCUBSP35.EXT" PAGE 78

This program uses the Clamped Cubic Spline Method to construct the clamped cubic spline interpolant S for the function f. The sample problem considers $f(x) = e^{2x}$ on the interval $[0, 1]$.

INPUT: (Select input option 2.) CCUBSP35.DTA, $n = 4$, $FPO = 2$, $FP1 = 2e^2$

BÉZIER CURVE "BEZIER36.EXT" PAGE 89

This program uses the Bézier Curve method to construct parametric curves to approximate given data. The sample program considers

$$
\begin{aligned}
(x_0, y_0) &= (0, 0) \\
(x_0^+, y_0^+) &= (1/4, 1/4) \\
(x_1, y_1) &= (1, 1) \\
(x_1^-, y_1^-) &= (1/2, 1/2) \\
(x_1^+, y_1^+) &= (-1/2, -1/2) \\
(x_2, y_2) &= (2, 2) \\
(x_2^-, y_2^-) &= (-1, -1)
\end{aligned}
$$

INPUT: BEZIER36.DTA, $n = 2$

PROGRAMS FOR CHAPTER 4

COMPOSITE SIMPSON'S RULE "CSIMPR41.EXT" PAGE 102

This program uses Composite Simpson's Rule to approximate

$$\int_a^b f(x)dx.$$

The sample problem uses

$$f(x) = \sin x, \quad \text{on} \quad [0, \pi].$$

INPUT: $a = 0, \quad b = \pi, \quad n = 10$

ROMBERG INTEGRATION "ROMBRG42.EXT" PAGE 117

This program uses the Romberg Method to approximate

$$\int_a^b f(x)dx.$$

The sample problem uses

$$f(x) = \sin x, \quad \text{on} \quad [0, \pi].$$

INPUT: $a = 0, \quad b = \pi, \quad n = 6$

ADAPTIVE QUADRATURE "ADAPQR43.EXT" PAGE 122

This program uses the Adaptive Quadrature Method to approximate

$$\int_a^b f(x)dx$$

within a given tolerance $TOL > 0$. The sample problem uses

$$f(x) = \frac{100}{x^2}\sin\frac{10}{x}, \quad \text{on} \quad [1,3].$$

INPUT: $a = 1, \quad b = 3, \quad TOL = 0.0001, \quad N = 20$

COMPOSITE SIMPSON'S RULE FOR DOUBLE INTEGRALS "DINTGL44.EXT" PAGE 130

This program uses the Composite Simpson's Rule for Double Integrals to approximate

$$\int_a^b \int_{c(x)}^{d(x)} f(x,y)\ dydx.$$

The sample problem uses

$$f(x,y) = e^{\frac{y}{x}}$$

with

$$c(x) = x^3, \quad d(x) = x^2, \quad a = 0.1 \quad \text{and} \quad b = 0.5.$$

INPUT: $a = 0.1, \quad b = 0.5, \quad m = 5, \quad n = 5$

GAUSSIAN QAUDRATURE FOR DOUBLE INTEGRALS "DGQINT45.EXT" PAGE 130

This program uses Gaussian Quadrature to approximate

$$\int_a^b \int_{c(x)}^{d(x)} f(x,y)\ dydx.$$

The sample problem uses $f(x,y) = e^{y/x}$ with $c(x) = x^3, \quad d(x) = x^2, \quad a = 0.1$ and $b = 0.5$.

INPUT: $a = 0.1, \quad b = 0.5, \quad m = 5, \quad n = 5$

GAUSSIAN QUADRATURE FOR TRIPLE INTEGRALS "TINTGL46.EXT" PAGE 130

This program uses the Gaussian Quadrature to approximate

$$\int_a^b \int_{c(x)}^{d(x)} \int_{\alpha(x,y)}^{\beta(x,y)} f(x,y,z)\, dz dy dx.$$

The sample problem uses

$$f(x,y,z) = \sqrt{x^2+y^2}$$

with

$$\alpha(x,y) = \sqrt{x^2+y^2}, \quad \beta(x,y) = 2,$$

$$c(x) = 0.0, \quad d(x) = \sqrt{4-x^2}, \quad a = 0, \quad \text{and} \quad b = 2.$$

INPUT: $a = 0, \quad b = 2, \quad m = 5, \quad n = 5, \quad p = 5$

PROGRAMS FOR CHAPTER 5

EULER METHOD "EULERM51.EXT" PAGE 153

This program uses the Euler Method to approximate the solution of an initial value problem of the form

$$y' = f(t,y), \quad y(a) = \alpha, \quad a \le t \le b.$$

The sample problem uses

$$f(t,y) = y - t^2 + 1, \quad y(0) = 0.5, \quad 0 \le t \le 2.$$

INPUT: $a = 0, \quad b = 2, \quad N = 10, \quad \alpha = 0.5$

RUNGE-KUTTA METHOD OF ORDER FOUR "RKOR4M52.EXT" PAGE 165

This program uses the Runge-Kutta Method of order four to approximate the solution of the initial value problem of the form

$$y' = f(t, y), \qquad y(a) = \alpha, \qquad a \leq t \leq b.$$

The sample problem uses

$$f(t, y) = y - t^2 + 1, \qquad y(0) = 0.5, \qquad 0 \leq t \leq 2.$$

INPUT: $a = 0, \quad b = 2, \quad N = 10, \quad \alpha = 0.5$

ADAMS FOURTH-ORDER PREDICTOR-CORRECTOR METHOD "PRCORM53.EXT" PAGE 173

This program uses the Adams Fourth-Order Predictor-Corrector Method to approximate the solution of an initial value problem of the form

$$y' = f(t, y), \qquad y(a) = \alpha, \qquad a \leq t \leq b.$$

The sample problem uses

$$f(t, y) = y - t^2 + 1, \qquad y(0) = 0.5, \qquad 0 \leq t \leq 2.$$

INPUT: $a = 0, \quad b = 2, \quad N = 10, \quad \alpha = 0.5$

EXTRAPOLATION METHOD "EXTRAP54.EXT" PAGE 177

This program uses the Extrapolation Method to approximate the solution of an initial value problem of the form

$$y' = f(t, y), \qquad y(a) = \alpha, \qquad a \leq t \leq b.$$

The sample problem uses

$$f(t, y) = y - t^2 + 1, \qquad y(0) = 0.5, \qquad 0 \leq t \leq 2.$$

INPUT: $a = 0, \quad b = 2, \quad \alpha = 0.5, \quad TOL = 0.00001, \quad HMIN = 0.01,$
$HMAX = 0.25$

RUNGE-KUTTA-FEHLBERG METHOD "RKFVSM55.EXT" PAGE 183

This program uses the Runge-Kutta-Fehlberg Method to approximate the solution of the initial value problem of the form

$$y' = f(t, y), \qquad y(a) = \alpha, \qquad a \le t \le b$$

to within a given tolerance. The sample problem uses

$$f(t, y) = y - t^2 + 1, \qquad y(0) = 0.5, \qquad 0 \le t \le 2.$$

INPUT: $a = 0$, $b = 2$, $\alpha = 0.5$, $TOL = 0.00001$, $HMIN = 0.01$, $HMAX = 0.25$

ADAMS VARIABLE-STEPSIZE PREDICTOR-CORRECTOR METHOD "VPRCOR56.EXT" PAGE 185

This program uses the Adams Variable Stepsize Predictor-Corrector Method to approximate the solution of an initial value problem of the form

$$y' = f(t, y), \qquad y(a) = \alpha, \qquad a \le t \le b$$

given tolerance. The sample problem uses

$$f(t, y) = y - t^2 + 1, \qquad y(0) = 0.5, \qquad 0 \le t \le 2.$$

INPUT: $a = 0$, $b = 2$, $\alpha = 0.5$, $TOL = 0.00001$, $HMIN = 0.01$, $HMAX = 0.25$

RUNGE-KUTTA METHOD FOR SYSTEMS OF DIFFERENTIAL EQUATIONS "RKO4SY57.EXT" PAGE 189

This program uses the Runge-Kutta for Systems of Differential Equations Method to approximate a the solution of the mth-order system of first-order initial value problems.

The sample problem considers the second order system

$$f_1(u_1, u_2) = -4u_1 + 3u_2 + 6, \quad u_1(0) = 0$$

$$f_2(u_1, u_2) = -2.4u_1 + 1.6u_2 + 3.6, \quad u_2(0) = 0.$$

INPUT: $a = 0, \quad b = 0.5, \quad N = 5, \quad \alpha_1 = 0, \quad \alpha_2 = 0$

TRAPEZOIDAL METHOD WITH NEWTON ITERATION "TRAPNT58.EXT" PAGE 197

This program uses the Trapezoidal Method with Newton Iteration to approximate the solution to the initial value problem

$$y' = f(t, y), \qquad y(a) = \alpha, \qquad a \leq t \leq b.$$

The sample problem uses

$$f(t, y) = 5e^{5t}(y - t)^2 + 1, \qquad y(0) = -1, \qquad 0 \leq t \leq 1.$$

INPUT: $a = 0, \quad b = 1, \quad N = 5, \quad \alpha = -1, \quad TOL = 0.000001, \quad M = 10$

PROGRAMS FOR CHAPTER 6

GAUSSIAN ELIMINATION WITH BACKWARD SUBSTITUTION "GAUSEL61.EXT" PAGE 207

This program uses the Gaussian Elimination with Backward Substitution Method to solve an $n \times n$ linear system of the form $A\mathbf{x} = \mathbf{b}$. The sample problem solves the linear system

$$\begin{aligned} x_1 - x_2 + 2x_3 - x_4 &= -8 \\ 2x_1 - 2x_2 + 3x_3 - 3x_4 &= -20 \\ x_1 + x_2 + x_3 \quad &= -2 \\ x_1 - x_2 + 4x_3 + 3x_4 &= 4. \end{aligned}$$

INPUT: GAUSEL61.DTA, $n = 4$

GAUSSIAN ELIMINATION WITH MAXIMAL COLUMN PIVOTING "GAUMCP62.EXT" PAGE 213

This program uses the Gaussian Elimination with Maximal Column Pivoting Method to solve an $n \times n$ linear system. The sample problem solves the linear system

$$\begin{aligned} x_1 - x_2 + 2x_3 - x_4 &= -8 \\ 2x_1 - 2x_2 + 3x_3 - 3x_4 &= -20 \\ x_1 + x_2 + x_3 &= -2 \\ x_1 - x_2 + 4x_3 + 3x_4 &= 4. \end{aligned}$$

INPUT: GAUMCP62.DTA, $n = 4$

GAUSSIAN ELIMINATION WITH SCALED COLUMN PIVOTING "GAUSCP63.EXT" PAGE 214

This program uses the Gaussian Elimination with Scaled Column Pivoting Method to solve an $n \times n$ linear system. The sample problem solves the linear system

$$\begin{aligned} x_1 - x_2 + 2x_3 - x_4 &= -8 \\ 2x_1 - 2x_2 + 3x_3 - 3x_4 &= -20 \\ x_1 + x_2 + x_3 &= -2 \\ x_1 - x_2 + 4x_3 + 3x_4 &= 4. \end{aligned}$$

INPUT: GAUSCP63.DTA, $n = 4$

DIRECT FACTORIZATION "DIFACT64.EXT" PAGE 227

This program uses the Direct Factorization Method to factor the $n \times n$ matrix A into the product $A = LU$ of a lower triangular matrix L and an upper triangular matrix U. The matrix factored in the sample problem is

$$A = \begin{bmatrix} 6 & 2 & 1 & -1 \\ 2 & 4 & 1 & 0 \\ 1 & 1 & 4 & -1 \\ -1 & 0 & -1 & 3 \end{bmatrix}$$

INPUT: DIFACT64.DTA, $n = 4$, $ISW = 1$

CHOLESKI'S METHOD "CHOLFC65.EXT" PAGE 232

This program uses the Choleski Method to factor the positive definite $n \times n$ matrix A into the product LL^t, where L is a lower triangular matrix. The matrix factored in the sample problem is

$$A = \begin{bmatrix} 4 & -1 & 1 \\ -1 & 4.25 & 2.75 \\ 1 & 2.75 & 3.5 \end{bmatrix}$$

INPUT: CHOLFC65.DTA, $n = 3$

LDL^t FACTORIZATION "LDLFCT66.EXT" PAGE 232

This program uses the LDL^t Factorization Method to factor the positive definite $n \times n$ matrix A into the product LDL^t, where L is a lower triangular matrix and D is a diagonal matrix. The matrix factored in the sample problem is

$$A = \begin{bmatrix} 4 & -1 & 1 \\ -1 & 4.25 & 2.75 \\ 1 & 2.75 & 3.5 \end{bmatrix}$$

INPUT: LDLFCT66.DTA, $n = 3$

CROUT REDUCTION FOR TRIDIAGONAL LINEAR SYSTEMS "CRTRLS67.EXT" PAGE 235

This program uses the Crout Reduction for Tridiagonal Linear Systems Method to solve a tridiagonal $n \times n$ linear system. The sample system is

$$\begin{aligned} 2x_1 - x_2 \qquad\qquad &= 1 \\ -x_1 + 2x_2 - x_3 \qquad &= 0 \\ - x_2 + 2x_3 - x_4 &= 0 \\ - x_3 + 2x_4 &= 1. \end{aligned}$$

INPUT: CRTRLS67.DTA, $n = 4$

PROGRAMS FOR CHAPTER 7

JACOBI ITERATIVE METHOD "JACITR71.EXT" PAGE 258

This program uses the Jacobi Iterative Method to approximate the solution to the $n \times n$ linear system $A\mathbf{x} = \mathbf{b}$, given an initial approximation $\mathbf{x}_0 = (x_1^0, x_2^0, ..., x_n^0)^t$. The sample problem approximates the solution to the linear system

$$\begin{aligned} 10x_1 - x_2 + 2x_3 \quad &= 6 \\ -x_1 + 11x_2 - x_3 + 3x_4 &= 25 \\ 2x_1 - x_2 + 10x_3 - x_4 &= -11 \\ 3x_2 - x_3 + 8x_4 &= 15. \end{aligned}$$

starting with the initial vector $\mathbf{x}_0 = (0, 0, 0, 0)^t$.

INPUT: JACITR71.DTA, $n = 4$, $TOL = 0.001$, $N = 30$

GAUSS–SEIDEL ITERATIVE METHOD "GSEITR72.EXT" PAGE 259

This program uses the Gauss-Seidel Iterative Method to approximate the solution to the $n \times n$ linear system $A\mathbf{x} = \mathbf{b}$, given an initial approximation $\mathbf{x}_0 = (x_1^0, x_2^0, ..., x_n^0)^t$. The sample problem approximates the solution to the linear system

$$\begin{aligned} 10x_1 - x_2 + 2x_3 \quad &= 6 \\ -x_1 + 11x_2 - x_3 + 3x_4 &= 25 \\ 2x_1 - x_2 + 10x_3 - x_4 &= -11 \\ 3x_2 - x_3 + 8x_4 &= 15 \end{aligned}$$

starting with the initial vector $\mathbf{x}_0 = (0, 0, 0, 0)^t$.

INPUT: GSEITR72.DTA, $n = 4$, $TOL = 0.001$, $N = 30$

SUCCESSIVE–OVER–RELAXATION (SOR) METHOD "SORITR73.EXT" PAGE 261

This program uses the Successive-Over-Relaxation Method to approximate the solution to the $n \times n$ linear system $A\mathbf{x} = \mathbf{b}$, given a parameter ω and an initial approximation $\mathbf{x}_0 = (x_1^0, x_2^0, ..., x_n^0)^t$.

The sample problem approximates the solution to the linear system

$$\begin{aligned} 4x_1+3x_2 \quad\quad &= 24 \\ 3x_1+4x_2- x_3 &= 30 \\ - x_2+4x_3 &= -24 \end{aligned}$$

starting with the initial vector $\mathbf{x}_0 = (1,1,1)^t$.

INPUT: SORITR73.DTA, $n = 3$, $TOL = 0.001$, $N = 30$, $\omega = 1.25$

PROGRAMS FOR CHAPTER 8

PADÉ APPROXIMATION "PADEMD81.EXT" PAGE 292

This program uses Padé Approximation to compute the rational approximation

$$r(x) = \frac{p_0 + p_1 x + \cdots + p_n x^n}{q_0 + q_1 x + \cdots + q_m x^m}$$

to a function $f(x)$ given its Maclaurin series $a_0 + a_1 x + a_2 x^2 + \cdots$. The sample problem uses $f(x) = e^{-x}$, where $a_0 = 1$, $a_1 = -1$, $a_2 = \frac{1}{2}$, $a_3 = -\frac{1}{6}$, $a_4 = \frac{1}{24}$, $a_5 = -\frac{1}{120}$.

INPUT: PADEMD81.DTA, $m = 2$, $n = 3$

CHEBYSHEV RATIONAL APPROXIMATION "CHEBYM82.EXT" PAGE 295

This program uses Chebyshev Rational Approximation to compute the rational approximation

$$r_T(x) = \frac{p_0 T_0(x) + p_1 T_1(x) + \cdots + p_n T_n(x)}{q_0 T_0(x) + q_1 T_1(x) + \cdots + q_m T_m(x)}$$

to a function $f(x)$ given its Chebyshev expansion $a_0 T_0(x) + a_1 T_1(x) + a_2 T_2(x) + \cdots$. The sample problem uses $f(x) = e^{-x}$, where $a_0 = 1.266066$, $a_1 = -1.130318$, $a_2 = 0.271495$, $a_3 = -0.044337$, $a_4 = 0.005474$, $a_5 = -0.000543$.

INPUT: CHEBYM82.DTA, $m = 2$, $n = 3$

FAST FOURIER TRANSFORM METHOD "FFTRNS83.EXT" PAGE 309

This program uses the Fast Fourier Transform Method to compute the coefficients in the discrete trigonometric approximation for a given set of data. The sample problem constructs an approximation to the function

$$f(x) = e^{-x}$$

on the interval [0,2].

INPUT: (Select input option 3.) $m = 8$

PROGRAMS FOR CHAPTER 9

POWER METHOD "POWERM91.EXT" PAGE 320

This program uses the Power Method to approximate the dominant eigenvalue and an associated eigenvector of an $n \times n$ matrix A given a nonzero vector $\mathbf{x}$. The sample problem considers the matrix

$$A = \begin{bmatrix} -4 & 14 & 0 \\ -5 & 13 & 0 \\ -1 & 0 & 2 \end{bmatrix}$$

with $\mathbf{x} = (1,1,1)^t$ as the initial approximation to the eigenvector.

INPUT: POWERM91.DTA, $n = 3$, $TOL = 0.0001$, $N = 30$

SYMMETRIC POWER METHOD "SYMPWR92.EXT" PAGE 321

This program uses the Symmetric Power Method to approximate the dominant eigenvalue and an associated eigenvector of a symmetric $n \times n$ matrix A given a nonzero vector $\mathbf{x}$. The sample problem considers the symmetric matrix

$$A = \begin{bmatrix} 4 & -1 & 1 \\ -1 & 3 & -2 \\ 1 & -2 & 3 \end{bmatrix}$$

with $\mathbf{x} = (1,0,0)^t$ as the initial approximation to the eigenvector.

INPUT: SYMPWR92.DTA, $n = 3$, $TOL = 0.0001$, $N = 25$

INVERSE POWER METHOD "INVPWR93.EXT" PAGE 324

This program uses the Inverse Power Method to approximate an eigenvalue nearest to a given number q and an associated eigenvector of an $n \times n$ matrix A. The sample problem considers the matrix

$$A = \begin{bmatrix} -4 & 14 & 0 \\ -5 & 13 & 0 \\ -1 & 0 & 2 \end{bmatrix}$$

with $\mathbf{x} = (1,1,1)^t$ as the initial approximation to the eigenvector and the number q defined by

$$q = \frac{\mathbf{x}^{(0)t} A \mathbf{x}^{(0)}}{\mathbf{x}^{(0)t}\mathbf{x}^{(0)}}.$$

INPUT: INVPWR93.DTA, $n = 3$, $TOL = 0.0001$, $N = 25$

WIELANDT DEFLATION "WIEDEF94.EXT" PAGE 325

This program uses the Wielandt Deflation Method to approximate the second most dominant eigenvalue and an associated eigenvector of the $n \times n$ matrix A given a nonzero vector $\mathbf{x}_0$. The sample problem considers the matrix

$$A = \begin{bmatrix} 4 & -1 & 1 \\ -1 & 3 & -2 \\ 1 & -2 & 3 \end{bmatrix}$$

which has the dominant eigenvalue $\lambda = 6$ and associated eigenvector $\mathbf{v} = (1,-1,1)^t$. The initial approximation $\mathbf{x}_0 = (1,1)^t$.

INPUT: WIEDEF94.DTA, $n = 3$, $TOL = 0.0001$, $N = 30$

HOUSEHOLDER'S METHOD "HSEHLD95.EXT" PAGE 333

This program uses the Householder Method to obtain a symmetric tridiagonal matrix that is similiar to a given symmetric matrix A. The sample problem considers the matrix

$$A = \begin{bmatrix} 4 & 1 & -2 & 2 \\ 1 & 2 & 0 & 1 \\ -2 & 0 & 3 & -2 \\ 2 & 1 & -2 & -1 \end{bmatrix}$$

INPUT: HSEHLD95.DTA, $n = 4$

QR METHOD "QRSYMT96.EXT" PAGE 338

This program uses the QR Method to obtain the eigenvalues of a symmetric, tridiagonal $n \times n$ matrix of the form

$$A = \begin{bmatrix} a_1^{(1)} & b_2^{(1)} & 0 & \cdots & \cdots & \cdots & \cdots & 0 \\ b_2^{(1)} & a_2^{(1)} & b_3^{(1)} & \ddots & & & & \vdots \\ 0 & b_3^{(1)} & a_3^{(1)} & b_4^{(1)} & \ddots & & & \vdots \\ \vdots & \ddots & \ddots & \ddots & \ddots & \ddots & & \vdots \\ \vdots & & \ddots & \ddots & \ddots & \ddots & \ddots & \vdots \\ \vdots & & & \ddots & \ddots & \ddots & \ddots & 0 \\ \vdots & & & & \ddots & b_{n-1}^{(1)} & a_{n-1}^{(1)} & b_n^{(1)} \\ 0 & \cdots & \cdots & \cdots & \cdots & 0 & b_n^{(1)} & a_n^{(1)} \end{bmatrix}$$

The sample problem considers the matrix

$$A = \begin{bmatrix} a_1^{(1)} & b_2^{(1)} & 0 \\ b_2^{(1)} & a_2^{(1)} & b_3^{(1)} \\ 0 & b_3^{(1)} & a_3^{(1)} \end{bmatrix} = \begin{bmatrix} 3 & 1 & 0 \\ 1 & 3 & 1 \\ 0 & 1 & 3 \end{bmatrix}$$

INPUT: QRSYMT96.DTA, $n = 3$, $TOL = 0.00001$, $M = 30$

PROGRAMS FOR CHAPTER 10

NEWTON'S METHOD FOR SYSTEMS "NWTSY101.EXT" PAGE 347

This program uses Newton's Method for Systems to approximate the solution of the nonlinear system of equations $\mathbf{F}(x) = \mathbf{0}$ given an initial approximation $\mathbf{x}_0$. The sample problem uses

$$\mathbf{F}(x) = (f_1(\mathbf{x}), f_2(\mathbf{x}), f_3(\mathbf{x}))^t$$

where

$$\mathbf{x} = (x_1, x_2, x_3)^t$$

and

$$f_1(x_1, x_2, x_3) = 3x_1 - \cos(x_2 x_3) - 0.5$$
$$f_2(x_1, x_2, x_3) = x_1^2 - 81(x_2 + 0.1)^2 + \sin x_3 + 1.06$$
$$f_3(x_1, x_2, x_3) = e^{-x_1 x_2} + 20x_3 + \frac{10\pi - 3}{3}.$$

INPUT: $n = 3, \quad TOL = 0.00001, \quad N = 25, \quad \mathbf{x}_0 = (0.1, 0.1, -0.1)^t$

BROYDEN'S METHOD "BROYM102.EXT" PAGE 354

This program uses the Broyden Method to approximate the solution of the nonlinear system of equations $\mathbf{F}(x) = \mathbf{0}$ given an initial approximation $\mathbf{x}_0$. The sample problem uses

$$\mathbf{F}(x) = (f_1(\mathbf{x}), f_2(\mathbf{x}), f_3(\mathbf{x}))^t$$

where

$$\mathbf{x} = (x_1, x_2, x_3)^t$$

and

$$f_1(x_1, x_2, x_3) = 3x_1 - \cos(x_2 x_3) - 0.5$$
$$f_2(x_1, x_2, x_3) = x_1^2 - 81(x_2 + 0.1)^2 + \sin x_3 + 1.06$$
$$f_3(x_1, x_2, x_3) = e^{-x_1 x_2} + 20x_3 + \frac{10\pi - 3}{3}.$$

INPUT: $n = 3, \quad TOL = 0.00001, \quad N = 25, \quad \mathbf{x}_0 = (0.1, 0.1, -0.1)^t$

STEEPEST DESCENT METHOD "STPDC103.EXT" PAGE 359

This program uses the Steepest Descent Method to approximate a solution to the minimum of the function

$$g(\mathbf{x}) = \sum_{i=1}^{n} [f_i(\mathbf{x})]^2$$

given an initial approximation $\mathbf{x}_0$. This also approximates a zero of

$$\mathbf{F}(x) = (f_1(\mathbf{x}), f_2(\mathbf{x}), \ldots, f_n(\mathbf{x}))^t.$$

The sample problem uses

$$\mathbf{F}(x) = (f_1(\mathbf{x}), f_2(\mathbf{x}), f_3(\mathbf{x}))^t \quad \text{where} \quad \mathbf{x} = (x_1, x_2, x_3)^t$$

and

$$f_1(x_1, x_2, x_3) = 3x_1 - \cos(x_2 x_3) - 0.5$$
$$f_2(x_1, x_2, x_3) = x_1^2 - 81(x_2 + 0.1)^2 + \sin x_3 + 1.06$$
$$f_3(x_1, x_2, x_3) = e^{-x_1 x_2} + 20x_3 + \frac{10\pi - 3}{3}.$$

INPUT: $n = 3, \quad TOL = 0.05, \quad N = 10, \quad \mathbf{x}_0 = (0.5, 0.5, 0.5)^t$

PROGRAMS FOR CHAPTER 11

LINEAR SHOOTING METHOD "LINST111.EXT" PAGE 366

This program uses the Linear Shooting Method to approximate the solution of a linear two-point boundary-value problem

$$y'' = p(x)y' + q(x)y + r(x), \quad a \leq x \leq b, \quad y(a) = \alpha, \quad y(b) = \beta.$$

The sample problem considers the boundary-value problem

$$y'' = -\frac{2}{x}y' + \frac{2}{x^2}y + \frac{\sin(\ln x)}{x^2}, \quad y(1) = 1, \quad y(2) = 2.$$

INPUT: $a = 1, \quad b = 2, \quad \alpha = 1, \quad \beta = 2, \quad N = 10$

LINEAR FINITE-DIFFERENCE METHOD "LINFD112.EXT" PAGE 372

This program uses the Linear Finite-Difference Method to approximate the solution of a linear two-point boundary-value problem

$$y'' = p(x)y' + q(x)y + r(x), \quad a \leq x \leq b, \quad y(a) = \alpha, \quad y(b) = \beta.$$

The sample problem considers the boundary-value problem

$$y'' = -\frac{2}{x}y' + \frac{2}{x^2}y + \frac{\sin(\ln x)}{x^2}, \quad y(1) = 1, \quad y(2) = 2.$$

INPUT: $a = 1, \quad b = 2, \quad \alpha = 1, \quad \beta = 2, \quad N = 9$

NONLINEAR SHOOTING METHOD "NLINS113.EXT" PAGE 378

This program uses the Nonlinear Shooting Method to approximate the solution of a nonlinear two-point boundary-value problem

$$y'' = f(x, y, y'), \quad a \le x \le b, \quad y(a) = \alpha, \quad y(b) = \beta.$$

The sample problem considers the boundary-value problem

$$y'' = 4 + 0.25x^3 - 0.125yy', \quad y(1) = 17, \quad y(3) = \frac{43}{3}$$

where

$$f_y(x, y, y') = -\frac{y'}{8} \quad \text{and} \quad f_{y'}(x, y, y') = -\frac{y}{8}.$$

INPUT: $a = 1$, $b = 3$, $\alpha = 17$, $\beta = \frac{43}{3}$, $N = 20$, $TOL = 0.0001$, $M = 25$

NONLINEAR FINITE-DIFFERENCE METHOD "NLFDM114.EXT" PAGE 380

This program uses the Nonlinear Finite-Difference Method to approximate the solution to a nonlinear two-point boundary-value problem

$$y'' = f(x, y, y'), \quad a \le x \le b, \quad y(a) = \alpha, \quad y(b) = \beta.$$

The sample problem considers the boundary-value problem

$$y'' = 4 + 0.25x^3 - 0.125yy', \quad y(1) = 17, \quad y(3) = \frac{43}{3}$$

where

$$f_y(x, y, y') = -\frac{y'}{8} \quad \text{and} \quad f_{y'}(x, y, y') = -\frac{y}{8}.$$

INPUT: $a = 1$, $b = 3$, $\alpha = 17$, $\beta = \frac{43}{3}$, $N = 19$, $TOL = 0.0001$, $M = 25$

PIECEWISE LINEAR RAYLEIGH-RITZ METHOD "PLRRG115.EXT" PAGE 387

This program uses the Piecewise Linear Rayleigh-Ritz Method to approximate the solution to a two-point boundary-value problem

$$-\frac{d}{dx}\left(p(x)\frac{dy}{dx}\right) + q(x)y = f(x), \quad 0 \le x \le 1, \quad y(0) = y(1) = 0.$$

The sample problem uses the differential equation

$$-y'' + \pi^2 y = 2\pi^2 \sin \pi x, \quad y(0) = 0, \quad y(1) = 0.$$

INPUT: $n = 9$, PLRRG115.DTA

CUBIC SPLINE "CSRRG116.EXT" PAGE 391
RAYLEIGH-RITZ METHOD

This program uses the Cubic Spline Rayleigh-Ritz Method to approximate the solution of a boundary-value problem

$$-\frac{d}{dx}\left(p(x)\frac{dy}{dx}\right)+q(x)y=f(x),\quad 0\le x\le 1,\quad y(0)=y(1)=0.$$

The sample problem uses the differential equation

$$-y''+\pi^2 y=2\pi^2\sin\pi x,\quad y(0)=0,\quad y(1)=0.$$

INPUT: $n=9,\quad f'(0)=2\pi^3,\quad f'(1)=-2\pi^3,\quad p'(0)=0,\quad p'(1)=0,$
$q'(0)=0,\quad q'(1)=0$

PROGRAMS FOR CHAPTER 12

POISSON EQUATION "POIFD121.EXT" PAGE 402
FINITE-DIFFERENCE METHOD

This program uses the Poisson Equation Finite-Difference Method to approximate the solution to the Poisson equation

$$\frac{\partial^2 u}{\partial x^2}(x,y)+\frac{\partial^2 u}{\partial y^2}(x,y)=f(x,y)$$

subject to boundary conditions $u(x,y)=g(x,y)$. The sample problem uses

$$f(x,y)=xe^y \quad\text{and}\quad g(x,y)=xe^y.$$

INPUT: $a=0,\quad b=2,\quad c=0,\quad d=1,\quad n=6,\quad m=5,$
$TOL=10^{-5},\quad M=150$

HEAT EQUATION "HEBDM122.EXT" PAGE 411
BACKWARD-DIFFERENCE METHOD

This program uses the Heat Equation Backward-Difference Method to approximate the solution to a parabolic partial-differential equation

$$\frac{\partial u}{\partial t}(x,t)=\alpha^2\frac{\partial^2 u}{\partial x^2}(x,t),\quad 0<x<l,\quad 0<t$$

subject to zero boundary conditions and the initial condition $u(x,0) = f(x)$. The sample problem uses

$$f(x) = \sin \pi x.$$

INPUT: $l = 1,\quad T = 0.5,\quad \alpha = 1,\quad m = 10,\quad N = 50$

CRANK-NICOLSON METHOD "HECNM123.EXT" PAGE 413

This program uses the Crank-Nicolson Method to approximate the solution to a parabolic partial-differential equation

$$\frac{\partial u}{\partial t}(x,t) = \alpha^2 \frac{\partial^2 u}{\partial x^2}(x,t), \quad 0 < x < l, \quad 0 < t$$

subject to zero boundary conditions and the initial condition $u(x,0) = f(x)$. The sample problem uses

$$f(x) = \sin \pi x.$$

INPUT: $l = 1,\quad T = 0.5,\quad \alpha = 1,\quad m = 10,\quad N = 50$

WAVE EQUATION "WVFDM124.EXT" PAGE 420
FINITE-DIFFERENCE METHOD

This program uses the Wave Equation Finite-Difference Method to approximate the solution to a wave equation

$$\frac{\partial^2 u}{\partial t^2}(x,t) - \alpha^2 \frac{\partial^2 u}{\partial x^2}(x,t) = 0, \quad 0 < x < l, \quad 0 < t$$

subject to zero boundary conditions and initial conditions

$$u(x,0) = f(x) \quad \text{and} \quad \frac{\partial u}{\partial t}(x,0) = g(x).$$

The sample problem uses

$$f(x) = \sin \pi x \quad \text{and} \quad g(x) = 0.$$

INPUT: $l = 1,\quad T = 1,\quad \alpha = 2,\quad m = 10,\quad N = 20$

FINITE-ELEMENT METHOD "LINFE125.EXT" PAGE 429

This program uses the Finite-Element Method to approximate the solution to an elliptic partial-differential equation of the form

$$\frac{\partial}{\partial x}\left(p(x,y)\frac{\partial u}{\partial x}\right)+\frac{\partial}{\partial y}\left(q(x,y)\frac{\partial u}{\partial y}\right)+r(x,y)u=f(x,y)$$

subject to Dirichlet, mixed, or Neumann boundary conditions. The sample problem considers Laplace's equation

$$\frac{\partial^2 u}{\partial x^2}(x,y)+\frac{\partial^2 u}{\partial y^2}(x,y)=0$$

on the two-dimensional region shown in the figure below.

The boundary conditions on this region are

$$u(x,y)=4 \quad \text{for } (x,y) \text{ on } L_6 \text{ and } L_7,$$

$$\frac{\partial u}{\partial n}(x,y)=x \quad \text{for } (x,y) \text{ on } L_2 \text{ and } L_4,$$

$$\frac{\partial u}{\partial n}(x,y)=y \quad \text{for } (x,y) \text{ on } L_5,$$

and

$$\frac{\partial u}{\partial n}(x,y)=\frac{x+y}{\sqrt{2}} \quad \text{for } (x,y) \text{ on } L_1 \text{ and } L_3.$$

INPUT: LINFE125.DTA

Chapter 1 Mathematical Preliminaries and Error Analysis

EXERCISE SET 1.2 (*Page* 9)

1. For each part, $f \in C[a,b]$ on the given interval. Since $f(a)$ and $f(b)$ are of opposite sign, the Intermediate Value Theorem implies a number c exists with $f(c) = 0$.

2. **a)** $[0,1]$ **b)** $[0,1]$, $[4,5]$, $[-1,0]$
c) $[-2,-1]$, $[0,1]$, $[2.5,3.5]$ **d)** $[-3,-2]$, $[-1,-0.5]$, and $[-0.5,0]$

3. The maximum value for $|f(x)|$ is given below.
a) 0.4620981 **b)** 1.333333 **c)** 5.164000
d) 1.582572 **e)** 160 **f)** 10.79869

4. **a)** $P_3(x) = -4 + 6x - x^2 - 4x^3$; $P_3(0.4) = -2.016$
b) $|R_3(0.4)| \leq 0.05849$; $|f(0.4) - P_3(0.4)| = 0.013365367$
c) $P_4(x) = -4 + 6x - x^2 - 4x^3$; $P_4(0.4) = -2.016$
d) $|R_4(0.4)| \leq 0.01366$; $|f(0.4) - P_4(0.4)| = 0.013365367$

5. **a)** $P_3(x) = -1 + 4x - 2x^2 - \frac{1}{2}x^3$ and $P_3(0.25) = -0.1328125$
b) $|R_3(0.25)| \leq 2.005 \times 10^{-4}$, $|f(0.25) - P_3(0.25)| = 4.061 \times 10^{-5}$
c) $P_4(x) = -1 + 4x - 2x^2 - \frac{1}{2}x^3$ and $P_4(0.25) = -0.1328125$
d) $|R_4(0.25)| \leq 4.069 \times 10^{-5}$, $|f(0.25) - P_4(0.25)| = 4.061 \times 10^{-5}$

6. $P_5(x) = (x-1) - (x-1)^2/2 + (x-1)^3/3 - (x-1)^4/4 + (x-1)^5/5$; $|R_5(x)| \leq 1.215 \times 10^{-4}$

7. $P_4(x) = 1 + 2x + \frac{3x^2}{2} + \frac{x^3}{3} - \frac{7x^4}{24}$; $|P_4(x)| \leq 0.03753$

8. $P_3(x) = \sqrt{3}/2 + \frac{1}{2}(x-\pi/3) - \sqrt{3}/4(x-\pi/3)^2 - \frac{1}{12}(x-\pi/3)^3$ satisfies $|\sin 58° - P_3(1.0122909)| \leq 5.4 \times 10^{-8}$

9. $P_3(x) = \frac{\sqrt{3}}{2} - \frac{1}{2}\left(x - \frac{\pi}{6}\right) - \frac{\sqrt{3}}{4}\left(x - \frac{\pi}{6}\right)^2 + \frac{1}{12}\left(x - \frac{\pi}{6}\right)^3$ and $|\cos 32° - P_3(0.5585054)| \leq 5.4 \times 10^{-8}$.

10. For n odd, $P_n(x) = x - \frac{1}{3}x^3 + \frac{1}{5}x^5 + \cdots + (-1)^{\frac{n-1}{2}} x^n/n$;
For n even, $P_n(x) = P_{n-1}(x)$.

11. **a)** Expanding about $x_0 = 0$ we have $P_1(x) = 7 - 2x$, $P_2(x) = 7 - 2x$, $P_3(x) = 7 - 2x + 5x^3$, and $P_4(x) = 7 - 2x + 5x^3$.

b) Expanding about $x_0 = 1$ we have $P_1(x) = 10 + 13(x-1)$, $P_2(x) = 10 + 13(x-1) + 15(x-1)^2$, $P_3(x) = 10 + 13(x-1) + 15(x-1)^2 + 5(x-1)^3$, and $P_4(x) = 10 + 13(x-1) + 15(x-1)^2 + 5(x-1)^3$.

12. A bound for the maximum error is 0.00261.

13. 0.176

EXERCISE SET 1.3 (*Page* 14)

1. The machine numbers are equivalent to

a) 2707 **b)** −2707 **c)** 0.0174560546875 **d)** 0.0174560584130

2.

	Next Largest	Next Smallest
a)	2707.00024414	2706.99975586
b)	−2706.99975586	2707.00024414
c)	0.0174560584130	0.0174560509622
d)	0.0174560621381	0.0174560546875

3.

	Absolute Error	Relative Error
a)	0.001264	4.025×10^{-4}
b)	7.346×10^{-6}	2.338×10^{-6}
c)	2.818×10^{-4}	1.037×10^{-4}
d)	2.136×10^{-4}	1.510×10^{-4}
e)	2.647×10^{1}	1.202×10^{-3}
f)	1.454×10^{1}	1.050×10^{-2}
g)	420	1.042×10^{-2}
h)	3.343×10^{3}	9.213×10^{-3}

4. The calculations and their errors are:

a)	(i) 17/15	(ii) 1.13	(iii) 1.13	(iv) both 3×10^{-3}
b)	(i) 4/15	(ii) 0.266	(iii) 0.266	(iv) both 2.5×10^{-3}
c)	(i) 139/660	(ii) 0.211	(iii) 0.210	(iv) 2×10^{-3}, 3×10^{-3}
d)	(i) 310/660	(ii) 0.455	(iii) 0.456	(iv) 2×10^{-3}, 1×10^{-4}

5.

	Approximation	Absolute Error	Relative Error
a)	134	0.079	5.90×10^{-4}
b)	133	0.499	3.77×10^{-3}
c)	2.00	0.327	0.195
d)	1.67	0.003	1.79×10^{-3}
e)	1.80	0.154	0.0786
f)	−15.1	0.0546	3.60×10^{-3}
g)	0.286	2.86×10^{-4}	10^{-3}
h)	0.00	0.0215	1.00

6.

	Approximation	Absolute Error	Relative Error
a)	3.14557613	3.983×10^{-3}	1.268×10^{-3}
b)	3.14162103	2.838×10^{-5}	9.032×10^{-6}

7.

	Approximation	Absolute Error	Relative Error
a)	133.9	0.021	1.568×10^{-4}
b)	132.5	0.001	7.55×10^{-6}
c)	1.700	0.027	0.01614
d)	1.673	0	0
e)	1.986	0.03246	0.01662
f)	−15.16	0.005377	3.548×10^{-4}
g)	0.2857	1.429×10^{-5}	5×10^{-5}
h)	−0.01700	0.0045	0.2092

8.

	Approximation	Absolute Error	Round-off Error
a)	133	0.921	6.88×10^{-3}
b)	132	0.501	3.78×10^{-3}
c)	1.00	0.673	0.402
d)	1.67	0.003	1.79×10^{-3}
e)	3.55	1.59	0.817
f)	−15.2	0.0453	0.00299
g)	0.284	0.00171	0.00600
h)	0	0.02149	1

9.

	Approximation	Absolute Error	Relative Error
a)	133.9	0.021	1.568×10^{-4}
b)	132.5	0.001	7.55×10^{-6}
c)	1.600	0.073	0.04363
d)	1.673	0	0
e)	1.983	0.02945	0.01508
f)	−15.15	0.004622	3.050×10^{-4}
g)	0.2855	2.142×10^{-4}	7.5×10^{-4}
h)	−0.01700	0.0045	0.2092

10.

	Approximation	Absolute Error	Relative Error
a)	3.145	0.003407	0.001085
b)	3.139	0.002593	8.253×10^{-4}

11.

	Approximation	Absolute Error	Relative Error
a)	3.145	0.003407	1.085×10^{-3}
b)	3.139	0.002593	8.253×10^{-4}

EXERCISE SET 1.4 (*Page* 19)

1.

	x_1	Absolute Error	Relative Error
a)	92.26	1.542×10^{-2}	1.672×10^{-4}
b)	0.005421	1.264×10^{-6}	2.333×10^{-4}
c)	10.98	6.874×10^{-3}	6.257×10^{-4}
d)	−0.001149	7.566×10^{-8}	6.584×10^{-5}

	x_2	Absolute Error	Relative Error
a)	0.005421	6.273×10^{-7}	1.157×10^{-4}
b)	−92.26	4.581×10^{-3}	4.966×10^{-5}
c)	0.001149	7.560×10^{-8}	6.580×10^{-5}
d)	−10.98	6.875×10^{-3}	6.257×10^{-4}

2.

	x_1	Absolute Error	Relative Error
a)	92.24	0.004580	4.965×10^{-5}
b)	0.005417	2.736×10^{-6}	5.048×10^{-4}
c)	10.98	6.874×10^{-3}	6.257×10^{-4}
d)	−0.001149	7.566×10^{-8}	6.584×10^{-5}

	x_2	Absolute Error	Relative Error
a)	0.005418	2.373×10^{-6}	4.377×10^{-4}
b)	−92.25	5.419×10^{-3}	5.874×10^{-5}
c)	0.00149	7.560×10^{-8}	6.580×10^{-5}
d)	−10.98	6.875×10^{-3}	6.257×10^{-4}

3. **a)** −0.1000 **b)** −0.1010

c) The absolute error for part (a) is 2.331×10^{-3} with relative error 2.387×10^{-2}. The absolute error for part (b) is 3.331×10^{-3} with relative error 3.411×10^{-2}.

4. **a)** −0.09000 **b)** −0.1140

c) The absolute error for (a) is 7.669×10^{-3}, and the relative error for (a) is 7.852×10^{-2}. The absolute error for (b) is 1.633×10^{-2}, and the relative error for (b) is 1.672×10^{-1}

5. The approximate solutions to the systems are

a) $x = 2.451,\ y = -1.635$ **b)** $x = 507.7,\ y = 82.00$

6. The approximate solutions to the systems are

a) $x = 2.460 \quad y = -1.634$ **b)** $x = 477.0 \quad y = 76.93$

7.

	Approximation	Absolute Error	Relative Error
a)	3.743×10^{-1}	1.011×10^{-3}	2.694×10^{-3}
b)	3.755×10^{-1}	1.889×10^{-4}	5.033×10^{-4}

8.

	Approximation	Absolute Error	Relative Error
a)	0.3743	1.011×10^{-3}	2.694×10^{-3}
b)	0.3756	2.889×10^{-4}	7.698×10^{-4}

9. **a)** $O(\frac{1}{n})$ **b)** $O(\frac{1}{n^2})$ **c)** $O(\frac{1}{n})$ **d)** $O(\frac{1}{n})$

10. **a)** $O(h^2)$ **b)** $O(h)$ **c)** $O(h^2)$ **d)** $O(h)$

CHAPTER 2 SOLUTIONS OF EQUATIONS OF ONE VARIABLE

EXERCISE SET 2.2 (*Page* 29)

1. The Bisection method gives

a) $p_7 = 0.5859$ **b)** $p_8 = 3.002$ **c)** $p_7 = 3.419$

2. The Bisection method gives

a) $p_7 = -1.414$ **b)** $p_8 = 1.414$ **c)** $p_7 = 2.727$ **d)** $p_7 = -0.7265$

3. The Bisection method gives $p_9 = 4.4932$, accurate to within 10^{-3}.

4. The Bisection method gives $p_{10} = 1.0068$, accurate to within 10^{-3}.

5. The Bisection method gives the follow approximations, accurate to within 10^{-5}:

a) $p_{17} = 0.641182$ **b)** $p_{17} = 0.257530$

c) On $[-3,-2]$, $p_{17} = -2.19307$; and on $[-1,0]$, $p_{17} = -0.798164$.

d) On $[0.2, 0.3]$, $p_{14} = 0.297528$; and on $[1.2, 1.3]$, $p_{14} = 1.256622$.

6. **a)** $p_7 = 0.711$ **b)** $p_7 = 1.180$

c) There is a solution in $[0.5, 1]$ and in $[1, 1.5]$. The method of testing determines which solution to seek.

7. A bound for the number of iterations is $n \geq 14$ and $p_{14} = 1.32477$.

8. A bound for the number of iterations is $n \geq 12$ and $p_{12} = 1.3787$.

9. A bound for the number of iterations is $n \geq 7$ and $p_7 = -0.473$.

10. A bound for the number of iterations is $n \geq 8$ and $p_8 = 0.708$.

11. The zeros of $f(x) = 4x\cos(2x) - (x-2)^2$ on $[0, 8]$ are given in the following table.

Interval	Approximation
$[0.1, 3]$	$p_{15} = 2.36324$
$[3, 4]$	$p_{14} = 3.81793$
$[4, 6]$	$p_{15} = 5.83929$
$[6, 8]$	$p_{15} = 6.60309$

12. The three smallest positive solutions of $\sin x = e^{-x}$ are given in the following table.

Interval	Approximation
$[0,2]$	$p_{11} = 0.5889$
$[2,5]$	$p_{12} = 3.0964$
$[5,7]$	$p_{11} = 6.2842$

13. For the interval $[0,1]$, $h = 0.1617$ so that the depth is $r - h = 0.838$ feet.

EXERCISE SET 2.3 (*Page* 34)

1. Using the endpoints of the intervals as p_0 and p_1, we have

a) $p_{11} = 2.69065$ **b)** $p_7 = -2.87939$ **c)** $p_6 = 0.73909$ **d)** $p_5 = 0.96433$

2. Using the endpoints of the intervals as p_0 and p_1, we have

a) $p_{16} = 2.69060$ **b)** $p_6 = -2.87938$ **c)** $p_7 = 0.73908$ **d)** $p_6 = 0.96433$

3. Using the endpoints of the intervals as p_0 and p_1, we have

a) $p_7 = 1.829384$ **b)** $p_9 = 1.397749$

c) $p_6 = 0.910008; p_{10} = 3.733079$ **d)** $p_8 = 1.412391; p_7 = 3.057104$

4. Using the endpoints of the intervals as p_0 and p_1, we have

a) $p_8 = 1.829383$ **b)** $p_{10} = 1.397751$

c) $p_6 = 2.370687; p_8 = 3.722112$ **d)** $p_{10} = 1.412392; p_{12} = 3.057099$

5. Using the endpoints of the intervals as p_0 and p_1, we have

a) $p_2 = -0.479$ **b)** $p_3 = 0.705$

6. a) $p^{(3)} = -0.479$ **b)** $p^{(4)} = 0.705$

7. Using the endpoints of the intervals as p_0 and p_1, we have

a) For $[2,3]$, $p_6 = 2.370687$ and for $[3,4]$, $p_7 = 3.722113$.

b) For $[0.2, 0.3]$, $p_4 = 0.297530$ and for $[1.2, 1.3]$, $p_5 = 1.256623$.

8. a) $p_{42} = -1.4139838$ **b)** $p_{44} = 0.6411829$

9. Using the endpoints of the intervals as p_0 and p_1, we have

a) $p_{5610} = -1.3977446$ **b)** $p_{3960} = 0.6491861$

10. Secant Method:
For $p_0 = 0.1$ and $p_1 = 3$ we have $p_7 = 2.363171$.
For $p_0 = 3$ and $p_1 = 4$ we have $p_7 = 3.817926$.
For $p_0 = 5$ and $p_1 = 6$ we have $p_6 = 5.839252$.
For $p_0 = 6$ and $p_1 = 7$ we have $p_9 = 6.603085$.

Newton's Method:
For $p_0 = 1.5$ we have $p_6 = 2.363171$ and for $p_0 = 3.5$ we have $p_5 = 3.817926$.
For $p_0 = 5.5$ we have $p_4 = 5.839252$ and for $p_0 = 7$ we have $p_5 = 6.603085$.

Method of False Position:
For $p_0 = 0.1$ and $p_1 = 3$ we have $p_{10} = 2.363171$.
For $p_0 = 3$ and $p_1 = 4$ we have $p_7 = 3.817926$.
For $p_0 = 5$ and $p_1 = 6$ we have $p_5 = 5.839252$.
For $p_0 = 6$ and $p_1 = 7$ we have $p_{13} = 6.603083$.

11. Secant Method:
For $p_0 = -4$ and $p_1 = -3$ we have $p_7 = -3.161950$.
For $p_0 = -3$ and $p_1 = -1$ we have $p_9 = -1.968873$.
For $p_0 = -1$ and $p_1 = 2$ we have $p_6 = 1.968873$.
For $p_0 = 3$ and $p_1 = 4$ we have $p_7 = 3.161950$.

Newton's Method:
For $p_0 = -3$ we have $p_4 = -3.161950$ and for $p_0 = -2$ we have $p_3 = -1.968873$.
For $p_0 = 1$ we have $p_4 = 1.968873$ and for $p_0 = 3$ we have $p_4 = 3.161950$.

Method of False Position:
For $p_0 = -4$ and $p_1 = -3$ we have $p_{13} = -3.161944$.
For $p_0 = -3$ and $p_1 = -1$ we have $p_{11} = -1.968875$.
For $p_0 = -1$ and $p_1 = 2$ we have $p_6 = 1.968873$.
For $p_0 = 3$ and $p_1 = 4$ we have $p_{13} = 3.161944$.

12. **a)** For $p_0 = 0.1$ and $p_1 = 1$ we have $p_{14} = 0.2324$.

b) For $p_0 = 0.1$ and $p_1 = 1$ we have $p_8 = 0.2324$.

c) For $p_0 = 0.1$ and $p_1 = 1$ we have $p_{88} = 0.2324$.

13. $w_4 = -0.3170618$

EXERCISE SET 2.4 (*Page* 39)

1. **a)** For $p_0 = 2$ we have $p_5 = 2.69065$.
b) For $p_0 = -3$ we have $p_3 = -2.87939$.
c) For $p_0 = 0$ we have $p_4 = 0.73909$.
d) For $p_0 = 0.0$ we have $p_3 = 0.96434$.

2. **a)** For $p_0 = 1$ we have $p_8 = 1.829384$.
b) For $p_0 = 1.5$ we have $p_4 = 1.397748$.
c) For $p_0 = 1$ we have $p_4 = 0.910008$ and for $p_0 = 3$ we have $p_9 = 3.733079$.
d) For $p_0 = 1$ we have $p_4 = 1.412391$ and for $p_0 = 4$ we have $p_5 = 3.057104$.

3. Newton's method gives the approximation $p_3 = 0.90479$.

4. **a)** $p_0 = 1,\ p_3 = 1.007624$

b) $p_0 = 2.5,\ p_3 = 2.370687;\ p_0 = 3.5,\ p_4 = 3.722113$

c) $p_0 = -3,\ p_6 = -2.191308;\ p_0 = -1,\ p_4 = -0.798160$

d) $p_0 = 1,\ p_4 = 1.412391;\ p_0 = 4,\ p_5 = 3.057104$

5. Secant Method:
For $p_0 = 0.1$ and $p_1 = 3$ we have $p_7 = 2.363171$.
For $p_0 = 3$ and $p_1 = 4$ we have $p_7 = 3.817926$.
For $p_0 = 5$ and $p_1 = 6$ we have $p_6 = 5.839252$.
For $p_0 = 6$ and $p_1 = 7$ we have $p_9 = 6.603085$.

Newton's Method:
For $p_0 = 1.5$ we have $p_6 = 2.363171$ and for $p_0 = 3.5$ we have $p_5 = 3.817926$.
For $p_0 = 5.5$ we have $p_4 = 5.839252$ and for $p_0 = 7$ we have $p_5 = 6.603085$.

Method of False Position:
For $p_0 = 0.1$ and $p_1 = 3$ we have $p_{10} = 2.363171$.
For $p_0 = 3$ and $p_1 = 4$ we have $p_7 = 3.817926$.
For $p_0 = 5$ and $p_1 = 6$ we have $p_5 = 5.839252$.
For $p_0 = 6$ and $p_1 = 7$ we have $p_{13} = 6.603083$.

6. Secant Method:
For $p_0 = -4$ and $p_1 = -3$ we have $p_7 = -3.161950$.
For $p_0 = -3$ and $p_1 = -1$ we have $p_9 = -1.968873$.
For $p_0 = -1$ and $p_1 = 2$ we have $p_6 = 1.968873$.
For $p_0 = 3$ and $p_1 = 4$ we have $p_7 = 3.161950$.

Newton's Method:
For $p_0 = -3$ we have $p_4 = -3.161950$ and for $p_0 = -2$ we have $p_3 = -1.968873$.
For $p_0 = 1$ we have $p_4 = 1.968873$ and for $p_0 = 3$ we have $p_4 = 3.161950$.

Method of False Position:
For $p_0 = -4$ and $p_1 = -3$ we have $p_{13} = -3.161944$.
For $p_0 = -3$ and $p_1 = -1$ we have $p_{11} = -1.968875$.
For $p_0 = -1$ and $p_1 = 2$ we have $p_6 = 1.968873$.
For $p_0 = 3$ and $p_1 = 4$ we have $p_{13} = 3.161944$.

7. With $p_0 = \frac{\pi}{2}$, we have $p_{15} = 1.895488$; $p_0 = 5\pi$ gives $p_{19} = 1.895489$; and $p_0 = 10\pi$ does not converge in 200 iterations.

8. Newton's method gives the following:

a) For $p_0 = 0.5$ we have $p_{13} = 0.567135$.

b) For $p_0 = -1.5$ we have $p_{23} = -1.414325$.

c) For $p_0 = 0.5$ we have $p_{22} = 0.641166$.

d) For $p_0 = -0.5$ we have $p_{23} = -0.183274$.

9. **a)** For $p_0 = -1$ and $p_1 = 0$, we have $p_{17} = -0.04065850$; and for $p_0 = 0$ and $p_1 = 1$ we have $p_9 = 0.9623984$.

b) For $p_0 = -1$ and $p_1 = 0$, we have $p_5 = -0.04065929$; and for $p_0 = 0$ and $p_1 = 1$ we have $p_{12} = -0.04065929$.

c) For $p_0 = -0.5$, we have $p_5 = -0.04065929$; and for $p_0 = 0.5$ we have $p_{21} = 0.9623989$.

10. Modified Newton's method gives the following:

a) For $p_0 = 0.5$ we have $p_4 = 0.567143$.

b) For $p_0 = -1.5$ we have $p_2 = -1.414242$.

c) For $p_0 = 0.5$ we have $p_3 = 0.641186$.

d) For $p_0 = -0.5$ we have $p_5 = -0.183256$.

11. **a)** For $p_0 = -0.5$ we have $p_3 = -0.4341431$.

b) For $p_0 = 0.5$ we have $p_3 = 0.4506567$.
For $p_0 = 1.5$ we have $p_3 = 1.7447381$.
For $p_0 = 2.5$ we have $p_5 = 2.2383198$.
For $p_0 = 3.5$ we have $p_4 = 3.7090412$.

c) A reasonable initial approximation for the nth positive root is $n - 0.5$.

d) For $p_0 = 24.5$ we have $p_2 = 24.4998870$.

12. The two numbers are approximately 6.512849 and 13.487151.

13. 10.3%

14. **a)** $\frac{1}{3}e, t = 3$ hours **b)** 11 hours and 5 minutes **c)** 21 hours and 14 minutes

EXERCISE SET 2.5 (*Page* 44)

1. The results are recorded in the following table.

	a)	b)	c)	d)
$\hat{p}_0$	0.258684	0.907859	0.548101	0.731385
$\hat{p}_1$	0.257613	0.909568	0.547915	0.736087
$\hat{p}_2$	0.257536	0.909917	0.547847	0.737653
$\hat{p}_3$	0.257531	0.909989	0.547823	0.738469
$\hat{p}_4$	0.257530	0.910004	0.547814	0.738798
$\hat{p}_5$	0.257530	0.910007	0.547810	0.738958

2. Newton's Method gives $p_{18} = -0.183093$ and $\hat{p}_7 = -0.183387$.

3. $p_{10} = -0.169607$ and $\hat{p}_8 = -0.169607$ are accurate to within 2×10^{-4}

4. **a)** $p_0 = 0.5,\ \hat{p}_2 = 0.56717$

b) $p_0 = -1.5,\ \hat{p}_1 = -1.41422$

c) $p_0 = 0.5,\ \hat{p}_4 = 0.64124$

d) $p_0 = -0.5,\ \hat{p}_8 = -0.90279$

5. Aitken's Δ^2 method gives

a) $\hat{p}_{10} = 0.0\overline{45}$ **b)** $\hat{p}_2 = 0.0363$

6. For $k > 0, \lim_{n\to\infty} \frac{|p_{n+1}-0|}{|p_n-0|} = \lim_{n\to\infty} \frac{\frac{1}{(n+1)^k}}{\frac{1}{n^k}} = \lim_{n\to\infty} \left(\frac{n}{n+1}\right)^k = 1$, so the convergence is linear. We will have $1/N^k < 10^{-m}$ when $N > 10^{m/k}$.

7. a) Since

$$\lim_{n\to\infty} \frac{|p_{n+1}-0|}{|p_n-0|^2} = \lim_{n\to\infty} \frac{10^{-2^{n+1}}}{(10^{-2^n})^2} = \lim_{n\to\infty} \frac{10^{-2^{n+1}}}{10^{-2^{n+1}}} = 1,$$

the sequence is quadratically convergent.

b) Since

$$\lim_{n\to\infty} \frac{|p_{n+1}-0|}{|p_n-0|^2} = \lim_{n\to\infty} \frac{10^{-(n+1)^k}}{(10^{-n^k})^2} = \lim_{n\to\infty} \frac{10^{-(n+1)^k}}{10^{-2n^k}} = \lim_{n\to\infty} 10^{2n^k-(n+1)^k}$$

diverges, the sequence $p_n = 10^{-n^k}$ does not converge quadratically.

EXERCISE SET 2.6 (*Page* 48)

1. a) For $p_0 = 1$ we have $p_{22} = 2.69065$.

b) For $p_0 = 1$ we have $p_5 = 0.53209$, for $p_0 = -1$ we have $p_3 = -0.65270$, and for $p_0 = -3$ we have $p_3 = -2.87939$.

c) For $p_0 = 1$ we have $p_5 = 1.32472$.

d) For $p_0 = 1$ we have $p_4 = 1.12412$, and for $p_0 = 0$ we have $p_8 = -0.87605$.

e) For $p_0 = 0$ we have $p_6 = -0.47006$, for $p_0 = -1$ we have $p_4 = -0.88533$, and for $p_0 = -3$ we have $p_4 = -2.64561$.

f) For $p_0 = 0$ we have $p_{10} = 1.49819$.

2. a) For $p_0 = 0$ we have $p_9 = -4.123106$ and for $p_0 = 3$ we have $p_6 = 4.123106$. The complex roots are $-2.5 \pm 1.322879i$.

b) For $p_0 = 1$ we have $p_7 = -3.548233$ and for $p_0 = 4$ we have $p_5 = 4.38113$. The complex roots are $0.5835597 \pm 1.494188i$.

c) The only roots are complex and they are $\pm\sqrt{2}i$ and $-0.5 \pm 0.5\sqrt{3}i$.

d) For $p_0 = 1$ we have $p_5 = -0.250237$, for $p_0 = 2$ we have $p_5 = 2.260086$, and for $p_0 = -11$ we have $p_6 = -12.612430$. The complex roots are $-0.1987094 \pm 0.8133125i$.

e) For $p_0 = 0$ we have $p_8 = 0.846743$, and for $p_0 = -1$ we have $p_9 = -3.358044$. The complex roots are $-1.494350 \pm 1.744219i$.

f) For $p_0 = 0$ we have $p_8 = 2.069323$, and for $p_0 = 1$ we have $p_3 = 0.861174$. The complex roots are $-1.465248 \pm 0.8116722i$.

g) For $p_0 = 0$ we have $p_6 = -0.732051$, for $p_0 = 1$ we have $p_4 = 1.414214$, for $p_0 = -1$ we have $p_5 = -0.732051$, and for $p_0 = -2$ we have $p_6 = -1.414214$.

h) For $p_0 = 0$ we have $p_5 = 0.585786$, for $p_0 = 2$ we have $p_2 = 3$, and for $p_0 = 4$ we have $p_6 = 3.414214$.

3. The following table lists the initial approximation and the roots found by Müller's method.

	p_0	p_1	p_2	Approximate Roots	Complex Conjugate Roots
a)	-1	0	1	$p_7 = -0.34532 - 1.31873i$	$-0.34532 + 1.31873i$
	0	1	2	$p_6 = 2.69065$	
b)	0	1	2	$p_6 = 0.53209$	
	1	2	3	$p_9 = -0.65270$	
	-2	-3	-2.5	$p_4 = -2.87939$	
c)	0	1	2	$p_5 = 1.32472$	
	-2	-1	0	$p_7 = -0.66236 - 0.56228i$	$-0.66236 + 0.56228i$
d)	0	1	2	$p_5 = 1.12412$	
	2	3	4	$p_{12} = -0.12403 + 1.74096i$	$-0.12403 - 1.74096i$
	-2	0	-1	$p_5 = -0.87605$	
e)	0	1	2	$p_{10} = -0.88533$	
	1	0	-0.5	$p_5 = -0.47006$	
	-1	-2	-3	$p_5 = -2.64561$	
f)	0	1	2	$p_6 = 1.49819$	
	-1	-2	-3	$p_{10} = -0.51363 - 1.09156i$	$-0.51363 + 1.09156i$
	1	0	-1	$p_8 = 0.26454 - 1.32837i$	$0.26454 + 1.32837i$

4. The following table lists the initial approximations and the roots.

	p_0	p_1	p_2	Approximated Roots	Complex Conjugate Roots
a)	0	1	2	$p_{11} = -2.5 - 1.322876i$	$-2.5 + 1.322876i$
	1	2	3	$p_6 = 4.123106$	
	-3	-4	-5	$p_5 = -4.123106$	
b)	0	1	2	$p_7 = 0.583560 - 1.494188i$	$0.583560 + 1.494188i$
	2	3	4	$p_6 = 4.381113$	
	-2	-3	-4	$p_5 = -3.548233$	
c)	0	1	2	$p_{11} = 1.414214i$	$-1.414214i$
	-1	-2	-3	$p_{10} = -0.5 + 0.866025i$	$-0.5 - 0.866025i$
d)	0	1	2	$p_7 = 2.260086$	
	3	4	5	$p_{14} = -0.198710 + 0.813313i$	$-0.198710 + 0.813313i$
	11	12	13	$p_{22} = -0.250237$	
	-9	-10	-11	$p_6 = -12.612430$	
e)	0	1	2	$p_6 = 0.846743$	
	3	4	5	$p_{12} = -1.494349 + 1.744218i$	$-1.494349 - 1.744218i$
	-1	-2	-3	$p_7 = -3.358044$	
f)	0	1	2	$p_6 = 2.069323$	
	-1	0	1	$p_5 = 0.861174$	
	-1	-2	-3	$p_8 = -1.465248 + 0.811672i$	$-1.465248 - 0.811672i$
g)	0	1	2	$p_6 = 1.414214$	
	-2	-1	0	$p_7 = -0.732051$	
	0	-2	-1	$p_7 = -1.414214$	
	2	3	4	$p_6 = 2.732051$	
h)	0	1	2	$p_8 = 3$	
	-1	0	1	$p_5 = 0.585786$	
	2.5	3.5	4	$p_6 = 3.414214$	

5. **a)** The zeros are 1.244, 8.847 and -1.091, and the critical points are 0 and 6.

   b) The zeros are 0.5798, 1.521, 2.332, and -2.432, and the critical points are 1, 2.001, and -1.5.

6. Since the volume $V = 1000 = \pi r^2 h$, we have $h = 1000/(\pi r^2)$. The amount of material $M(r)$ is given by

$$M(r) = 2\pi(r + 0.25)^2 + (2\pi r + 0.25)h = 2\pi(r + 0.25)^2 + 2000/r + 250\pi/r^2.$$

   Thus,

$$M'(r) = 4\pi(r + 0.25) - 2000/r^2 - 500/(\pi r^3).$$

   Solving $M'(r) = 0$ for r gives $r = 5.363858$. Thus, $M(r) = 598.1813$ cm^2.

7. The width is approximately $W = 16.2121$ ft.

CHAPTER 3 INTERPOLATION AND POLYNOMIAL APPROXIMATION

EXERCISE SET 3.2 (*Page* 61)

1. a)

n	$x_0, x_1, \ldots, x_n$	$P_n(8.4)$
1	8.3, 8.6	17.87833
2	8.3, 8.6, 8.7	17.87716
3	8.1, 8.3, 8.6, 8.7	17.81000
4	8.0, 8.1, 8.3, 8.6, 8.7	17.60859

b)

n	$x_0, x_1, \ldots, x_n$	$P_n(-1/3)$
1	−0.5, −0.25	0.21504167
2	−0.5, −0.25, 0.0	0.16988889
3	−0.75, −0.5, −0.25, 0.0	0.17451852
4	−1.0, −0.75, −0.5, −0.25, 0.0	0.17451852

c)

n	$x_0, x_1, \ldots, x_n$	$P_n(0.25)$
1	0.2, 0.3	−0.13869287
2	0.2, 0.3, 0.4	−0.13259734
3	0.1, 0.2, 0.3, 0.4	−0.13277477
4	0.0, 0.1, 0.2, 0.3, 0.4	−0.13277246

d)

n	$x_0, x_1, \ldots, x_n$	$P_n(0.9)$
1	0.8, 1.0	0.44086280
2	0.7, 0.8, 1.0	0.43841352
3	0.6, 0.7, 0.8, 1.0	0.44198500
4	0.5, 0.6, 0.7, 0.8, 1.0	0.44325624

e)

n	$x_0, x_1, \ldots, x_n$	$P_n(\pi)$
1	3.1, 3.2	−3.122526
2	3.0, 3.1, 3.2	−3.141593
3	2.9, 3.0, 3.1, 3.2	−3.141686
4	2.9, 3.0, 3.1, 3.2, 3.4	−3.141590

f)

n	$x_0, x_1, \ldots, x_n$	$P_n(\pi/2)$
1	1.4, 1.5	19.98002
2	1.3, 1.4, 1.5	23.67269
3	1.2, 1.3, 1.4, 1.5	26.36966
4	1.1, 1.2, 1.3, 1.4, 1.5	28.49372

g)

n	$x_0, x_1, \ldots, x_n$	$P_n(1.15)$
1	1.1, 1.2	2.074637
2	1.0, 1.1, 1.2	2.076485
3	1.0, 1.1, 1.2, 1.3	2.076244
4	1.0, 1.1, 1.2, 1.3, 1.4	2.076216

h)

n	$x_0, x_1, \ldots, x_n$	$P_n(4.1)$
1	4.0, 4.2	−0.0262639
2	4.0, 4.2, 4.4	0.0005223
3	3.8, 4.0, 4.2, 4.4	−0.0022188
4	3.6, 3.8, 4.0, 4.2, 4.4	−0.0024334

2. a)

n	$x_0, x_1, \ldots, x_n$	$P_n(8.4)$
1	8.3, 8.6	17.87833
2	8.3, 8.6, 8.7	17.87716
3	8.1, 8.3, 8.6, 8.7	17.81000
4	8.0, 8.1, 8.3, 8.6, 8.7	17.60859

b)

n	$x_0, x_1, \ldots, x_n$	$P_n(-1/3)$
1	−0.5, −0.25	0.21504167
2	−0.5, −0.25, 0.0	0.16988889
3	−0.75, −0.5, −0.25, 0.0	0.17451852
4	−1.0, −0.75, −0.5, −0.25, 0.0	0.17451852

c)

n	$x_0, x_1, \ldots, x_n$	$P_n(0.25)$
1	0.2, 0.3	−0.13869287
2	0.2, 0.3, 0.4	−0.13259734
3	0.1, 0.2, 0.3, 0.4	−0.13277477
4	0.0, 0.1, 0.2, 0.3, 0.4	−0.13277246

d)

n	$x_0, x_1, \ldots, x_n$	$P_n(0.9)$
1	0.8, 1.0	0.44086280
2	0.7, 0.8, 1.0	0.43841352
3	0.6, 0.7, 0.8, 1.0	0.44198500
4	0.5, 0.6, 0.7, 0.8, 1.0	0.44325624

e)

n	$x_0, x_1, \ldots, x_n$	$P_n(\pi)$
1	3.1, 3.2	−3.122526
2	3.0, 3.1, 3.2	−3.141593
3	2.9, 3.0, 3.1, 3.2	−3.141686
4	2.9, 3.0, 3.1, 3.2, 3.4	−3.141590

f)

n	$x_0, x_1, \ldots, x_n$	$P_n(\pi/2)$
1	1.4, 1.5	19.98002
2	1.3, 1.4, 1.5	23.67269
3	1.2, 1.3, 1.4, 1.5	26.36966
4	1.1, 1.2, 1.3, 1.4, 1.5	28.49372

g)

n	$x_0, x_1, \ldots, x_n$	$P_n(1.15)$
1	1.1, 1.2	2.074637
2	1.0, 1.1, 1.2	2.076485
3	1.0, 1.1, 1.2, 1.3	2.076244
4	1.0, 1.1, 1.2, 1.3, 1.4	2.076216

h)

n	$x_0, x_1, \ldots, x_n$	$P_n(4.1)$
1	4.0, 4.2	−0.0262639
2	4.0, 4.2, 4.4	0.0005223
3	3.8, 4.0, 4.2, 4.4	−0.0022188
4	3.6, 3.8, 4.0, 4.2, 4.4	−0.0024334

3. We have $\sqrt{3} \approx P_4(1/2) = 1.708\overline{3}$.

4. We have $\sqrt{3} \approx P_4(3) = 1.690607$.

5. a)

n	Actual Error	Error Bound
1	0.00118	0.00120
2	1.367×10^{-5}	1.452×10^{-5}
3	3.830×10^{-6}	2.823×10^{-7}
4	3.470×10^{-6}	8.789×10^{-9}

b)

n	Actual Error	Error Bound
1	4.0523×10^{-2}	4.5153×10^{-2}
2	4.6296×10^{-3}	4.6296×10^{-3}
3	0	0
4	0	0

c)

n	Actual Error	Error Bound
1	5.9210×10^{-3}	6.0971×10^{-3}
2	1.7455×10^{-4}	1.8128×10^{-4}
3	2.8798×10^{-6}	4.5143×10^{-6}
4	5.6320×10^{-7}	5.8594×10^{-7}

d)

n	Actual Error	Error Bound
1	2.7296×10^{-3}	1.4080×10^{-2}
2	5.1789×10^{-3}	9.2215×10^{-3}

6. We have $f(1.09) \approx 0.2826$. The actual error is 4.3×10^{-5} and an error bound is 7.4×10^{-6}.

7. Neville's method with four-digit rounding arithmetic gives 0.2826.

8. The approximation is $\cos 0.75 \approx 0.7313$. The actual error is 0.0004 and an error bound is 2.7×10^{-8}. The discrepancy is due to the fact that round-off error occurs in the computation of the approximation.

9. a) $f(1.03) \approx P_{0,1,2} = 0.8094418$ b) $f(1.03) \approx P_{0,1,2,3} = 0.8092831$

10. **a)** 0.8094 **b)** 0.8090

11. **a)** $P_2(x) = -11.22017728x^2 + 3.808210517x + 1$ and an error bound is 0.11371294.

b) $P_2(x) = -0.1306344167x^2 + 0.8969979335x - 0.63249693$ and an error bound is 9.45762×10^{-4}.

c) $P_3(x) = 0.1970056667x^3 - 1.06259055x^2 + 2.532453189x - 1.666868305$ and an error bound is 10^{-4}.

d) $P_3(x) = -0.07932x^3 - 0.545506x^2 + 1.0065992x + 1$ and an error bound is 1.591376×10^{-3}.

12. **a)** 1.32436 **b)** 2.18350 **c)** 1.15277, 2.01191

d) Part (a) and (b) are better due to the spacing of the nodes.

13. The first ten terms of the sequence are 0.038462, 0.333671, 0.116605, −0.371760, −0.0548919, 0.605935, 0.190249, −0.513353, −0.0668173, and 0.448335. Since $f(1+\sqrt{10}) = 0.0545716$, the sequence does not appear to converge.

14. **a)** $P(1930) = 169,649,000$, $P(1965) = 191,767,000$, $P(2000) = 251,654,000$

EXERCISE SET 3.3 (*Page* 68)

1. **a)** $P_1(x) = 16.63553 + 9.7996(x-8)$; $P_1(8.4) = 20.55537$
$P_2(x) = P_1(x) - 33.50817(x-8)(x-8.1)$; $P_2(8.4) = 16.53439$
$P_3(x) = P_2(x) + 67.13678(x-8)(x-8.1)(x-8.3)$; $P_3(8.4) = 17.34003$
$P_4(x) = P_3(x) - 111.8981(x-8)(x-8.1)(x-8.3)(x-8.6)$; $P_4(8.4) = 17.60859$

b) $P_1(x) = -0.3440987 + 1.671541(x-0.5)$; $P_1(0.9) = 0.3245177$
$P_2(x) = P_1(x) + 1.177137(x-0.5)(x-0.6)$; $P_2(0.9) = 0.4657741$
$P_3(x) = P_2(x) - 0.7263767(x-0.5)(x-0.6)(x-0.7)$; $P_3(0.9) = 0.4483411$
$P_4(x) = P_3(x) - 2.118729(x-0.5)(x-0.6)(x-0.7)(x-0.8)$; $P_4(0.9) = 0.4432562$

c) $P_1(x) = -4.827866 + 5.87808(x-2.9)$; $P_1(\pi) = -3.407765$
$P_2(x) = P_1(x) + 7.76705(x-2.9)(x-3.0)$; $P_2(\pi) = -3.142072$
$P_3(x) = P_2(x) + 0.2711667(x-2.9)(x-3.0)(x-3.1)$; $P_3(\pi) = -3.141686$
$P_4(x) = P_3(x) - 1.159392(x-2.9)(x-3.0)(x-3.1)(x-3.2)$; $P_4(\pi) = -3.141590$

2. In the following equations we have $s = \frac{1}{h}(x - x_0)$.

a) $P_1(s) = 0.1 - 0.1718125s$; $P_1(-\frac{1}{3}) = -0.3581667$
$P_2(s) = P_1(s) + 0.218875s(s-1)/2$; $P_2(-\frac{1}{3}) = 0.1282222$
$P_3(s) = P_2(s) + 0.09375s(s-1)(s-2)/6$; $P_3(-\frac{1}{3}) = 0.1745185$
$P_4(s) = P_3(s)$; $P_4(-\frac{1}{3}) = 0.1745185$

b) $P_1(s) = -1 + 0.3795004s$; $P_1(0.25) = -0.05124895$
$P_2(s) = P_1(s) - 0.04298752s(s-1)/2$; $P_2(0.25) = -0.1318506$
$P_3(s) = P_2(s) - 2.93775 \times 10^{-3}s(s-1)(s-2)/6$; $P_3(0.25) = -0.1327686$
$P_4(s) = P_3(s) + 9.884 \times 10^{-5}s(s-1)(s-2)(s-3)/24$; $P_4(0.25) = -0.1327725$

c) $P_1(s) = 1.68437 + 0.265107s$; $P_1(1.15) = 2.082031$
$P_2(s) = P_1(s) - 0.014788s(s-1)/2$; $P_2(1.15) = 2.076485$
$P_3(s) = P_2(s) + 3.862 \times 10^{-3}s(s-1)(s-2)/6$; $P_3(1.15) = 2.076244$
$P_4(s) = P_3(s) - 1.194 \times 10^{-3}s(s-1)(s-2)(s-3)/24$; $P_4(1.15) = 2.076216$

3. In the following equations we have $s = \frac{1}{h}(x - x_n)$.

a) $P_1(s) = 1.101 + 0.7660625s$; $f(-\frac{1}{3}) \approx P_1(-\frac{4}{3}) = 0.07958333$
$P_2(s) = P_1(s) + 0.406375s(s+1)/2$; $f(-\frac{1}{3}) \approx P_2(-\frac{4}{3}) = 0.1698889$
$P_3(s) = P_2(s) + 0.09375s(s+1)(s+2)/6$; $f(-\frac{1}{3}) \approx P_3(-\frac{4}{3}) = 0.1745185$
$P_4(s) = P_3(s)$; $f(-\frac{1}{3}) \approx P_4(-\frac{4}{3}) = 0.1745185$

b) $P_1(s) = 0.2484244 + 0.2418235s$; $f(0.25) \approx P_1(-1.5) = -0.1143108$
$P_2(s) = P_1(s) - 0.04876419s(s+1)/2$; $f(0.25) \approx P_2(-1.5) = -0.1325973$
$P_3(s) = P_2(s) - 0.00283893s(s+1)(s+2)/6$; $f(0.25) \approx P_3(-1.5) = -0.1327748$
$P_4(s) = P_3(s) + 0.0009881s(s+1)(s+2)(s+3)/24$; $f(0.25) \approx P_4(-1.5) = -0.1327725$

c) $P_1(s) = 14.10142 + 8.303536s$; $f(\frac{\pi}{2}) \approx P_1(5\pi - 15) = 19.98002$
$P_2(s) = P_1(s) + 6.107754s(s+1)/2$; $f(\frac{\pi}{2}) \approx P_2(5\pi - 15) = 23.67269$
$P_3(s) = P_2(s) + 4.941922s(s+1)(s+2)/6$; $f(\frac{\pi}{2}) \approx P_3(5\pi - 15) = 26.36966$
$P_4(s) = P_3(s) + 4.198648s(s+1)(s+2)(s+3)/24$; $f(\frac{\pi}{2}) \approx P_4(5\pi - 15) = 28.49372$

4. a) $P_4(x) = -6 + 1.05170x + 0.57250x(x-0.1) + 0.21500x(x-0.1)(x-0.3) + 0.06301x(x-0.1)(x-0.3)(x-0.6)$

b) Add $0.01415945x(x-0.1)(x-0.3)(x-0.6)(x-1)$ to the answer in part (a).

5. a) $f(0.05) \approx 1.05126$ b) $f(0.65) \approx 1.9155505$ c) $f(0.43) \approx 1.53725$

6. a) $P(1930) = 169,649,000$ b) $P(2000) = 251,654,000$

EXERCISE SET 3.4 (Page 72)

1. The coefficients for the polynomials in divided-difference form are given in the following tables. For example, the polynomial in part (a) is

$$H_3(x) = 17.56492 + 1.116256(x-8.3) + 6.726147(x-8.3)^2 - 44.44647(x-8.3)^2(x-8.6).$$

(a)	(b)	(c)	(d)
17.56492	22.363362	−0.02475	−4.240058
1.116256	2.1691753	0.751	6.64986
6.726147	0.01558225	2.751	7.8163
−44.44647	−3.2177925	1	0.2733
			−1.128
			−0.18

(e)	(f)	(g)	(h)
−0.62049958	2.572152	1.684370	1.16164956
3.5850208	7.615964	2.742245	−1.5072822
−2.1989182	26.83536	−0.9117500	−1.4545692
−0.490447	99.21040	0.9499000	−0.56972300
0.037205	580.7080	−0.8815000	−0.17751125
0.040475	3400.600	0.8600000	−0.0446421875
−0.0025277777	48026.08	−0.8722222	−0.0097682292
0.0029629628	678365.5	1.037037	−0.0016232639
		0.04629630	−0.00071614579
		−15.85648	0.00021927433

2.

	x	approximation to $f(x)$	actual $f(x)$	error
a)	8.4	17.83270	17.877146	0.0444463
b)	0.9	0.44392477	0.44359244	3.33233×10^{-4}
c)	$-\frac{1}{3}$	0.1745185	0.17451852	1.852×10^{-8}
d)	π	−3.141593	−3.1415927	3.4641×10^{-7}
e)	0.25	−0.1327719	−0.13277189	5.428×10^{-9}
f)	$\frac{\pi}{2}$	52.73927	∞	∞
g)	1.15	2.079908	2.0762091	0.00369887
h)	4.1	−0.0024596486	−0.0024596463	2.26×10^{-9}

3. For 2(a) we have an error bound of 5.9×10^{-8} and for 2(e) we have an error bound of 2.7×10^{-13}. The error bound for 2(c) is 0 since $f^{(n)}(x) \equiv 0$ for $n > 3$.

4. **a)** $H(1.03) = 0.80932485$. The actual error is 1.24×10^{-6} and and error bound is 1.31×10^{-6}.

b) $H(1.03) = 0.80932362$. The actual error is 10^{-10} and an error bound is 3.86×10^{-10}.

5. **a)** We have $\sin 0.34 \approx H_5(0.34) = 0.33349$.

b) The formula gives an error bound of 3.05×10^{-14}, but the actual error is 2.91×10^{-6}. The discrepancy is due to the fact that the data is only given to 5 decimal places.

c) We have $\sin 0.34 \approx H_7(0.34) = 0.33350$. Although the error bound is now 5.4×10^{-20}, the inaccuracy of the given data dominates the calculations. This result is actually less accurate than the approximation in part (b), since $\sin 0.34 = 0.333487$.

6. $H_3(1.25) = 1.1690804$ with an error bound of 4.81×10^{-5} and $H_5(1.25) = 1.1690161$ with an error bound of 4.43×10^{-5}.

7. The Hermite polynomial generated from these data is

$$\begin{aligned}H_9(x) =& 75x + 0.222222x^2(x-3) - 0.0311111x^2(x-3)^2 \\ &- 0.00644444x^2(x-3)^2(x-5) + 0.00226389x^2(x-3)^2(x-5)^2 \\ &- 0.000913194x^2(x-3)^2(x-5)^2(x-8) + 0.000130527x^2(x-3)^2(x-5)^2(x-8)^2 \\ &- 0.0000202236x^2(x-3)^2(x-5)^2(x-8)^2(x-13).\end{aligned}$$

a) The Hermite polynomial predicts a position of $H_9(10) = 743$ ft and a speed of $H_9'(10) = 36$ ft/sec. Although the position approximation is reasonable, the low speed prediction is suspect.

b) To find the first time the speed exceeds 55 mi/hr, which is equivalent to $80.\bar{6}$ ft/sec, we solve for the smallest value of t in the equation $80.\bar{6} = H_9'(x)$. This gives $x \approx 5.9119932$.

c) The estimated maximum speed is $H_9'(6.5009714) = 81.0004518$ ft/sec ≈ 55.228 mi/hr.

EXERCISE SET 3.5 (*Page* 81)

1. The equations of the respective free cubic splines are given by

$$S(x) = S_i(x) = a_i + b_i(x-x_i) + c_i(x-x_i)^2 + d_i(x-x_i)^3,$$

for x in $[x_i, x_{i+1}]$ and the coefficients in the following tables.

a)

i	a_i	b_i	c_i	d_i
0	17.564920	3.13410000	0.00000000	0.00000000

b)

i	a_i	b_i	c_i	d_i
0	0.22363362	2.17229175	0.00000000	0.00000000

c)

i	a_i	b_i	c_i	d_i
0	−0.02475000	1.03237500	0.00000000	6.50200000
1	0.33493750	2.25150000	4.87650000	−6.50200000

d)

i	a_i	b_i	c_i	d_i
0	−4.24005800	7.03907000	0.00000000	39.242000
1	−3.49690900	8.21633000	11.77260000	−39.242000

e)

i	a_i	b_i	c_i	d_i
0	−0.62049958	3.45508693	0.00000000	−8.9957933
1	−0.28398668	3.18521313	−2.69873800	−0.94630333
2	0.00660095	2.61707643	−2.98262900	9.9420966

f)

i	a_i	b_i	c_i	d_i
0	2.57215200	11.26245066	0.00000000	−96.295066
1	3.60210200	8.37359866	−28.88852000	1647.3073
2	5.79788400	52.01511466	465.30368000	−1551.0122

g)

i	a_i	b_i	c_i	d_i
0	1.68437000	2.68435107	0.00000000	−3.32810714
1	1.94947700	2.58450785	−0.99843214	1.85253571
2	2.19979600	2.44039750	−0.44267142	−0.22003571
3	2.43918900	2.34526214	−0.50868214	1.69560714

h)

i	a_i	b_i	c_i	d_i
0	1.16164956	−1.65814692	0.00000000	−3.50122696
1	0.80201036	−2.07829415	−2.10073617	0.53954232
2	0.30663842	−2.85384355	−1.77701078	−2.99443232
3	−0.35916618	−3.92397974	−3.57367017	5.95611696

2.

	x	approximation to $f(x)$	actual $f(x)$	error
a)	8.4	17.87833	17.877146	1.1840×10^{-3}
b)	0.9	0.4408628	0.44359244	2.7296×10^{-3}
c)	$-\frac{1}{3}$	0.1774144	0.17451852	2.8959×10^{-3}
d)	π	−3.137628	−3.1415927	3.9647×10^{-3}
e)	0.25	−0.1315912	−0.13277189	1.1807×10^{-3}
f)	$\frac{\pi}{2}$	20.52772	∞	∞
g)	1.15	2.076438	2.0762091	2.289×10^{-4}
h)	4.1	4.895248×10^{-4}	−0.0024596463	2.949×10^{-3}

	x	approximation to $f'(x)$	actual $f'(x)$	actual error
a)	8.4	3.1341	3.128232	5.86829×10^{-3}
b)	0.9	2.172292	2.204367	0.0320747
c)	$-\frac{1}{3}$	1.574208	1.668000	0.093792
d)	π	8.991977	8.869604	0.1223726
e)	0.25	2.908242	2.907061	1.18057×10^{-3}
f)	$\frac{\pi}{2}$	94.57712	∞	∞
g)	1.15	2.498559	2.501619	3.05982×10^{-3}
h)	4.1	−3.299079	−3.319944	0.0208652

3. The equations of the respective clamped cubic splines are given by

$$s(x) = s_i(x) = a_i + b_i(x - x_i) + c_i(x - x_i)^2 + d_i(x - x_i)^3,$$

for x in $[x_i, x_{i+1}]$ and the coefficients in the following tables.

a)

i	a_i	b_i	c_i	d_i
0	17.564920	1.1162560	20.060086	−44.446466

b)

i	a_i	b_i	c_i	d_i
0	0.22363362	2.1691753	0.65914075	−3.2177925

c)

i	a_i	b_i	c_i	d_i
0	−0.02475000	0.75100000	2.5010000	1.0000000
1	0.33493750	2.18900000	3.2510000	1.0000000

d)

i	a_i	b_i	c_i	d_i
0	−4.2400580	6.6498600	7.7887900	0.27510000
1	−3.4969090	8.2158710	7.8713200	−0.18330000

e)

i	a_i	b_i	c_i	d_i
0	−0.62049958	3.5850208	−2.1498407	−0.49077413
1	−0.28398668	3.1403294	−2.2970730	−0.47458360
2	0.006600950	2.6666773	−2.4394481	−0.44980146

f)

i	a_i	b_i	c_i	d_i
0	2.5721520	7.6159640	−4.3305760	311.65936
1	3.6021020	16.099629	89.167232	−305.85328
2	5.7978840	24.757477	−2.5887520	5853.6757

g)

i	a_i	b_i	c_i	d_i
0	1.684370	2.742245	−1.005926	0.9417607
1	1.949477	2.569313	−0.723398	0.6217179
2	2.199796	2.443285	−0.536883	0.4333679
3	2.439189	2.348909	−0.406872	0.3128107

h)

i	a_i	b_i	c_i	d_i
0	1.16164956	−1.50728217	−1.34091868	−0.56825233
1	0.80201036	−2.11183992	−1.68187008	−0.71614399
2	0.30663842	−2.87052523	−2.11155648	−0.90466168
3	−0.35916618	−3.82370723	−2.65435349	−1.14727927

4.

	x	approximation to $f(x)$	actual $f(x)$	error
a)	8.4	17.83270	17.877146	0.044446
b)	0.9	0.4439248	0.44359244	3.3236×10^{-4}
c)	$-\frac{1}{3}$	0.174519	0.17451852	4.8×10^{-7}
d)	π	−3.141586	−3.1415927	6.7×10^{-6}
e)	0.25	−0.1327722	−0.13277189	3.1×10^{-7}
f)	$\frac{\pi}{2}$	39.11599	∞	∞
g)	1.15	2.072131	2.0762091	4.078×10^{-3}
h)	4.1	−0.00243433	−0.0024596463	2.532×10^{-5}

	x	approximation to $f'(x)$	actual $f'(x)$	error
a)	8.4	3.794879	3.128232	0.666647
b)	0.9	2.204470	2.204367	1.03286×10^{-4}
c)	$-\frac{1}{3}$	1.668000	1.668000	0.0
d)	π	8.869698	8.869604	9.35×10^{-5}
e)	0.25	2.907063	2.907061	4.319×10^{-7}
f)	$\frac{\pi}{2}$	112.4089	∞	∞
g)	1.15	2.501636	2.501619	1.7179×10^{-5}
h)	4.1	−3.319976	−3.319944	3.18169×10^{-5}

5. a) $S(x) = 0.29552 + 0.95430(x-0.3) - 4.5(x-0.3)^3$ on $[0.30, 0.32]$, and $S(x) = 0.31457 + 0.94973(x-0.32) - 0.27(x-0.32)^2 + 3(x-0.32)^3$ on $[0.32, 0.35]$; $S(0.34) = 0.33348$.

b) 7.09×10^{-6}

c) $s(x) = 0.29552 + 0.95433(x-0.3) - 0.1105(x-0.3)^2 - 1.5583(x-0.3)^3$ on $[0.30, 0.32]$;
$s(x) = 0.31457 + 0.94987(x-0.32) - 0.204(x-0.32)^2 + 0.64078(x-0.32)^3$ on $[0.32, 0.35]$;
$s(0.34) = 0.33349$.

d) 2.91×10^{-6} e) $S'(0.34) = 0.94253$ f) $s'(0.34) = 0.94248$

g) $\int_{0.30}^{0.35} S(x)\,dx = 0.015964$ h) $\int_{0.30}^{0.35} s(x)\,dx = 0.015964$

6. a) 0.33348 b) To 5 digits, $\sin 0.34 = 0.33349$

c) 0.33349 d) To 5 digits, $\sin 0.34 = 0.33349$

e) 0.94274 f) 0.94324

g) 0.015964 h) 0.015964

7. Using the portion of the spline on $[1.02, 1.04]$, we have $S(1.03) = 0.809324$ with an error bound of 1.9×10^{-7} and an actual error of 3.8×10^{-7}.

8. The piecewise linear approximation to f is given by

$$F(x) = \begin{cases} 20(e^{0.1}-1)x+1, & \text{for } x \text{ in } [0, 0.5] \\ 20(e^{0.2}-e^{0.1})x + 2e^{0.1} - e^{0.2}, & \text{for } x \text{ in } (0.5, 1]. \end{cases}$$

We have

$$\int_0^{0.1} F(x)\,dx = 0.1107936 \quad \text{and} \quad \int_0^{0.1} f(x)\,dx = 0.1107014.$$

9. On $[0, 0.05]$, $F(x) = 20(e^{0.1}-1)x+1$; and on $(0.05, 1]$, $F(x) = 20(e^{0.2}-e^{0.1})x + 2e^{0.1} - e^{0.2}$. So $\int_0^{0.1} F(x)\,dx = 0.1107936$.

10. $|f(x) - F(x)| \le \frac{M}{8} \max_{0 \le j \le n-1} |x_{j+1} - x_j|^2$, where $M = \max_{a \le x \le b} |f''(x)|$
Error bounds for Exercise 9 are 1.53×10^{-3} and 1.53×10^{-4}.

11. The spline has equation $S(x) = S_i(x) = a_i + b_i(x-x_i) + c_i(x-x_i)^2 + d_i(x-x_i)^3$ on $[x_i, x_{i+1}]$, where the coefficients are in the following table.

x_i	a_i	b_i	c_i	d_i
0	0	75	−0.659292	0.219764
3	225	76.99779	1.31858	−0.153761
5	383	80.4071	0.396018	−0.177237
8	623	77.9978	−1.19912	0.0799115

The spline predicts a position of $S(10) = 774.84$ ft and a speed of $S'(10) = 74.16$ ft/sec. To maximize the speed, we find the single critical point of $S'(x)$ and compare the values of $S(x)$ at this point and the endpoints. We find that max $S'(x) = S'(5.7448) = 80.7$ ft/sec $= 55.02$ mi/hr. The speed 55 mi/hr was first exceeded at approximately 5.5 sec.

12. The equation of the spline is

$$S(x) = S_i(x) = a_i + b_i(x - x_i) + c_i(x - x_i)^2 + d_i(x - x_i)^3$$

on the interval $[x_i, x_{i+1}]$, where the coefficients are given in the following table.

	Sample 1				Sample 2			
x_i	a_i	b_i	c_i	d_i	a_i	b_i	c_i	d_i
0	6.67	−0.44687	0	0.06176	6.67	1.6629	0	−0.00249
6	17.33	6.2237	1.1118	−0.27099	16.11	1.3943	−0.04477	−0.03251
10	42.67	2.1104	−2.1401	0.28109	18.89	−0.52442	−0.43490	0.05916
13	37.33	−3.1406	0.38974	−0.01411	15.00	−1.5365	0.09756	0.00226
17	30.10	−0.70021	0.22036	−0.02491	10.56	−0.64732	0.12473	−0.01113
20	29.31	−0.05069	−0.00386	0.00016	9.44	−0.19955	0.02453	−0.00102

13. **a)** $S(x) = S_i(x) = a_i + b_i(x - x_i) + c_i(x - x_i)^2 + d_i(x - x_i)^3$ on $[x_i, x_{i+1}]$, where

x_i	a_i	b_i	c_i	d_i
0	0	103.0425	0	−23.0820
0.25	25.4	98.7147	−17.3114	25.8098
0.5	49.4	94.8984	2.04590	1.91475
1.0	97.6	98.3803	4.91803	−6.55738
1.25				

b) 1:13 $\frac{7}{25}$

c) Starting speed $\approx 9.7047 \times 10^{-3}$ mi/sec $= 34.94$ mi/hr.
Ending speed ≈ 36.14 mi/hr.

14.

x_i	a_i	b_i	c_i	d_i
0	132165	1651.85	0.00000	2.64248
1	151326	2444.59	79.2744	−4.37641
2	179323	2717.16	−52.0179	2.00918
3	203302	2279.55	8.25746	−0.381311
4	226542	2330.31	−3.18186	0.106062

$S(1930) = 113004$, $S(1965) = 191860$, and $S(2000) = 272724$.

EXERCISE SET 3.6 (Page 89)

1. a) $x(t) = -10t^3 + 14t^2 + t, \quad y(t) = -2t^3 + 3t^2 + t$
   b) $x(t) = -10t^3 + 14.5t^2 + 0.5t, \quad y(t) = -3t^3 + 4.5t^2 + 0.5t$
   c) $x(t) = -10t^3 + 14t^2 + t, \quad y(t) = -4t^3 + 5t^2 + t$
   d) $x(t) = -10t^3 + 13t^2 + 2t, \quad y(t) = 2t$
2. a) $x(t) = -10t^3 + 12t^2 + 3t; \; y(t) = 2t^3 - 3t^2 + 3t$
   b) $x(t) = -10t^3 + 13.5t^2 + 1.5t; \; y(t) = -t^3 + 1.5t^2 + 1.5t$
   c) $x(t) = -10t^3 + 12t^2 + 3t; \; y(t) = -4t^3 + 3t^2 + 3t$
   d) $x(t) = -10t^3 + 9t^2 + 6t; \; y(t) = 8t^3 - 12t^2 + 6t$
3. a) $x(t) = -11.5t^3 + 15t^2 + 1.5t + 1, \quad y(t) = -4.25t^3 + 4.5t^2 + 0.75t + 1$
   b) $x(t) = -6.25t^3 + 10.5t^2 + 0.75t + 1, \quad y(t) = -3.5t^3 + 3t^2 + 1.5t + 1$
   c) Between $(0,0)$ and $(4,6)$ we have

$$x(t) = -5t^3 + 7.5t^2 + 1.5t, \quad y(t) = -13.5t^3 + 18t^2 + 1.5t,$$

   and between $(4,6)$ and $(6,1)$ we have

$$x(t) = -5.5t^3 + 6t^2 + 1.5t + 4, \quad y(t) = 4t^3 - 6t^2 - 3t + 6.$$

   d) Between $(0,0)$ and $(2,1)$ we have

$$x(t) = -5.5t^3 + 6t^2 + 1.5t, \quad y(t) = -1.25t^3 + 1.5t^2 + 0.75t,$$

   between $(2,1)$ and $(4,0)$ we have

$$x(t) = -4t^3 + 3t^2 + 3t + 2, \quad y(t) = -t^3 + 1,$$

   and between $(4,0)$ and $(6,-1)$ we have

$$x(t) = -8.5t^3 + 13.5t^2 - 3t + 4, \quad y(t) = -3.25t^3 + 5.25t^2 - 3t.$$

4. Between $(3,6)$ and $(2,2)$ we have

$$x(t) = 0.5t^3 - 2.4t^2 + 0.9t + 3, \quad y(t) = 6.5t^3 - 12t^2 + 1.5t + 6,$$

between $(2,2)$ and $(6,6)$ we have

$$x(t) = -5.9t^3 + 8.4t^2 + 1.5t + 2, \quad y(t) = -3.5t^3 + 6t^2 + 1.5t + 2,$$

between $(6,6)$ and $(5,2)$ we have

$$x(t) = -2.5t^3 + 4.5t^2 - 3t + 6, \quad y(t) = 6.8t^3 - 10.2t^2 - 0.6t + 6,$$

and between $(5,2)$ and $(6.5,3)$ we have

$$x(t) = -4.2t^3 + 7.2t^2 - 1.5t + 5, \quad y(t) = 0.1t^3 - 0.6t^2 + 1.5t + 2.$$

CHAPTER 4 NUMERICAL INTEGRATION AND DIFFERENTIATION

EXERCISE SET 4.2 (*Page* 98)

1. The Midpoint rule gives the following approximations:

a) 0.608197 **b)** 0.0 **c)** 0.8 **d)** 0.0

e) 2.08549 **f)** 0.5 **g)** 0.577350 **h)** 0.787179

2.

	Error Bound	Actual Error
a)	0.0416667	0.0280974
b)	866.981	19.9209
c)	0.0833333	0.0146018
d)	10.1670	6.28319
e)	235.120	16.2995
f)	0.0586667	0.0222443
g)	0.131183	0.0270256
h)	2.24625	0.173368

3. The Trapezoidal rule gives the following approximations:

a) 0.693147 **b)** 116.060 **c)** 0.75 **d)** −15.5031

e) −15.2556 **f)** 0.430769 **g)** 0.665431 **h)** 1.42209

4.

	Error Bound	Actual Error
a)	0.0833333	0.0568526
b)	1733.96	96.1391
c)	0.166667	0.0353982
d)	20.3339	9.21991
e)	470.240	1.04162
f)	0.117333	0.0469867
g)	0.262365	0.0610554
h)	4.49250	0.461543

5. Simpson's rule gives the following approximations:

a) 0.636514 **b)** 38.6865 **c)** 0.783333 **d)** −5.16771

e) −3.69486 **f)** 0.476923 **g)** 0.606711 **h)** 0.998816

6.

	Error Bound	Actual Error
a)	6.94444×10^{-4}	2.19639×10^{-4}
b)	399.339	18.7656
c)	8.33333×10^{-3}	2.06516×10^{-3}
d)	1.96009	1.11548
e)	49.7268	10.5191
f)	3.49867×10^{-3}	8.32723×10^{-4}
g)	0.0372082	2.33541×10^{-3}
h)	2.02491	0.0382688

7. The Midpoint rule gives the following approximations.

a) 0.174331 **b)** 0.151633 **c)** −0.176350 **d)** 0.0463500

e) 1.80391 **f)** −0.675325 **g)** 0.634335 **h)** 0.670379

8.

	Actual Error	Error Bound
a)	0.0179284	0.0198486
b)	8.96979×10^{-3}	0.0833333
c)	4.70020×10^{-4}	5.35341×10^{-4}
d)	0.0424053	0.0645858
e)	0.784719	1.14937
f)	0.0586442	0.280864
g)	1.87835×10^{-3}	3.35410×10^{-3}
h)	0.0276799	0.0403728

9. The Trapezoidal rule gives the following approximations.

a) 0.228074 **b)** 0.183940 **c)** −0.177764 **d)** 0.171287

e) 4.14326 **f)** −0.866667 **g)** 0.640046 **h)** 0.589049

10.

	Actual Error	Error Bound
a)	0.0358146	0.0396972
b)	0.0233372	0.1666667
c)	9.43980×10^{-4}	1.07068×10^{-3}
d)	0.0825317	0.129172
e)	1.55463	2.29874
f)	0.132698	0.561728
g)	3.83265×10^{-3}	6.70820×10^{-3}
h)	0.0536501	0.0807455

11. Simpson's rule gives the following approximations.

a) 0.192245 **b)** 0.162402 **c)** −0.176822 **d)** 0.0879957

e) 2.58370 **f)** −0.739105 **g)** 0.636239 **h)** 0.643269

12.

	Actual Error	Error Bound
a)	1.43597×10^{-5}	2.17014×10^{-5}
b)	1.79921×10^{-3}	4.16667×10^{-3}
c)	1.97990×10^{-6}	2.09475×10^{-6}
d)	7.59584×10^{-4}	1.29625×10^{-3}
e)	4.92863×10^{-3}	0.130283
f)	5.13583×10^{-3}	0.0632802
g)	2.56542×10^{-5}	6.70820×10^{-5}
h)	5.69918×10^{-4}	8.30132×10^{-4}

13.

	Approximation	Error Bound	Actual Error
a)	0.636394	3.08642×10^{-4}	9.94862×10^{-5}
b)	29.6524	177.484	9.73151
c)	0.784615	0.00370370	7.82775×10^{-4}
d)	−5.81368	0.871151	0.469508
e)	−9.39559	22.1008	4.81839
f)	0.477413	0.00155496	3.43065×10^{-4}
g)	0.605504	0.0165370	0.00112880
h)	0.980303	0.899961	0.0197555

14.

	Approximation	Error Bound	Actual Error
a)	0.6362962	1.24008×10^{-5}	1.84623×10^{-6}
b)	21.39650	23.7701	1.47565
c)	0.7855294	3.72024×10^{-4}	1.31249×10^{-4}
d)	−6.309332	0.0582514	0.0261471
e)	−14.12898	5.35642	0.0850017
f)	0.4777836	8.49795×10^{-5}	2.78918×10^{-5}
g)	0.6044737	0.543675×10^{-3}	9.80799×10^{-5}
h)	0.9636423	0.813615	3.09507×10^{-3}

15.

	Approximation	Error Bound	Actual Error
a)	0.617476	0.277778	0.0188183
b)	0.849966	577.988	19.0709
c)	0.796154	0.0555556	0.0107557
d)	−2.58386	6.77798	3.69933
e)	−7.44227	156.747	6.77171
f)	0.492960	0.0391111	0.0152047
g)	0.585529	0.0874551	0.0188469
h)	0.833041	1.49750	0.0127507

16.

	Approximation	Error Bound	Actual Error
a)	0.6361055	6.07639×10^{-4}	1.88870×10^{-4}
b)	6.267740	349.421	13.6531
c)	0.7874510	7.29167×10^{-3}	2.05282×10^{-3}
d)	−7.308250	1.71508	1.02506
e)	−23.25882	43.5109	9.04485
f)	0.4785366	3.06133×10^{-3}	7.80862×10^{-4}
g)	0.6025163	0.0325571	1.85930×10^{-3}
h)	0.9328652	1.77180	0.0276820

17.

	Approximation	Error Bound	Actual Error
a)	0.636162	4.22222×10^{-4}	1.32382×10^{-4}
b)	9.57271	242.798	10.3481
c)	0.786733	0.00506667	0.00133531
d)	−6.97811	1.19173	0.694927
e)	−20.6284	30.2339	6.41419
f)	0.478281	0.00212719	5.25342×10^{-4}
g)	0.603029	0.0226226	0.00134693
h)	0.939762	1.23115	0.0207852

EXERCISE SET 4.3 (*Page* 105)

1. The Composite Trapezoidal rule approximations are:

a) 0.639900	**b)** 31.3653	**c)** 0.784241	**d)** -6.42872
e) -13.5760	**f)** 0.476977	**g)** 0.605498	**h)** 0.970926

2. The Composite Midpoint rule approximations are:

a) 0.633096	**b)** 11.1568	**c)** 0.786700	**d)** -6.11274
e) -14.9985	**f)** 0.478751	**g)** 0.602961	**h)** 0.953892

3. The Composite Simpson's rule approximations are:

a) 0.6363098	**b)** 22.47713	**c)** 0.7853980	**d)** -6.2748768
e) -14.18334	**f)** 0.4777547	**g)** 0.6043941	**h)** 0.9610554

4. a) The Composite Trapezoidal rule approximation is 0.4215820.

b) The Composite Simpson's rule approximation is 0.4227162.

c) The Composite Midpoint rule approximation is 0.4241792.

5. a) The Composite Trapezoidal rule requires $h < 0.00092295$ and $n \geq 2167$.

b) The Composite Midpoint rule requires $h < 0.00065216$ and $n \geq 3066$.

c) The Composite Simpson's rule requires $h < 0.037658$ and $n \geq 54$.

6. a) The Composite Trapezoidal rule requires $h < 0.0022031$ and $n \geq 1426$.

b) The Composite Simpson's rule requires $h < 0.074653$ and $n \geq 44$.

c) The Composite Midpoint rule requires $h < 0.0015568$ and $n \geq 2016$.

7. a) The Composite Trapezoidal rule requires $h < 0.04382$ and $n \geq 46$. The approximation is 0.405471.

b) The Composite Midpoint rule requires $h < 0.03098$ and $n \geq 64$. The approximation is 0.405460.

c) The Composite Simpson's rule requires $h < 0.44267$ and $n \geq 6$. The approximation is 0.405466.

8. a) The Composite Trapezoidal rule requires $h < 0.01095$. With $n = 100$ the approximation is 0.6363001.

b) The Composite Simpson's rule requires $h < 0.173205$. With $n = 6$ the approximation is 0.6362975.

c) The Composite Midpoint rule requires $h < 0.0077460$. With $n = 130$ the approximation is 0.6362877.

9. The length is approximately 58.47047.

10. The Composite Trapezoidal rule gives the area of the bounded region as approximately 0.680164.

11. The Composite Simpson's rule gives the area of the bounded region as approximately 0.682711.

12. The length of the track is approximately 9858 ft.

13. **a)** Composite Simpson's rule with $h = 0.25$ gives 2.619719 seconds.

b) The Composite Trapezoidal rule with $h = 0.25$ gives 2.620866 seconds.

c) The exact result is

$$\int_{10}^{5} \frac{10}{-v\sqrt{v}}\, dv = 2.619716.$$

14. **a)** For $p_0 = 0.5$ we have $p_6 = 1.644854$ with $n = 20$.

b) For $p_0 = 0.5$ we have $p_6 = 1.645085$ with $n = 40$.

EXERCISE SET 4.4 (*Page* 111)

1. Gaussian quadrature with $n = 2$ gives

a) 0.1922687	**b)** 0.1594104	**c)** −0.1768190	**d)** 0.08926302
e) 2.5913247	**f)** −0.7307230	**g)** 0.6361966	**h)** 0.6423172

2. Gaussian quadrature with $n = 3$ gives

a) 0.1922594	**b)** 0.1605954	**c)** −0.1768200	**d)** 0.08875385
e) 2.589258	**f)** −0.7337990	**g)** 0.6362132	**h)** 0.7560233

3. Gaussian quadrature with $n = 4$ gives

a) 0.1922594	**b)** 0.1606028	**c)** −0.1768200	**d)** 0.08875529
e) 2.5886327	**f)** −0.7339604	**g)** 0.6362133	**h)** 0.6426991

4. Gaussian quadrature with $n = 5$ gives

a) 0.1922594	**b)** 0.1606028	**c)** −0.1768200	**d)** 0.08875528
e) 2.588629	**f)** −0.7339687	**g)** 0.6362133	**h)** 0.7560353

5. The Legendre polynomials P_2 and P_3 are given by

$$P_2(x) = \frac{1}{2}(3x^2 - 1) \quad \text{and} \quad P_3(x) = \frac{1}{2}(5x^3 - 3x),$$

so their roots are easily verified.
For $n = 2$,

$$c_1 = \int_{-1}^{1} \frac{x + 0.5773502692}{1.1547005}\,dx = 1$$

and

$$c_2 = \int_{-1}^{1} \frac{x - 0.5773502692}{-1.1547005}\,dx = 1.$$

For $n = 3$,

$$c_1 = \int_{-1}^{1} \frac{x(x + 0.7745966692)}{1.2}\,dx = \frac{5}{9},$$

$$c_2 = \int_{-1}^{1} \frac{(x + 0.7745966692)(x - 0.7745966692)}{-0.6}\,dx = \frac{8}{9},$$

and

$$c_3 = \int_{-1}^{1} \frac{x(x - 0.7745966692)}{1.2}\,dx = \frac{5}{9}.$$

6. For $n = 2$ we have 63.76955; for $n = 3$ we have 52.56146; for $n = 4$ we have 63.73053; and for $n = 5$ we have 52.98038.

EXERCISE SET 4.5 (*Page* 117)

1. Romberg integration gives $R_{3,3}$ as follows:

a) 0.1922593 **b)** 0.1606105 **c)** −0.1768200 **d)** 0.08875677
e) 2.5879685 **f)** −0.7341567 **g)** 0.6362135 **h)** 0.6426970

2. Romberg integration gives $R_{4,4}$ as follows:

a) 0.1922594 **b)** 0.1606028 **c)** −0.1768200 **d)** 0.08875528
e) 2.588627 **f)** −0.7339728 **g)** 0.6362134 **h)** 0.6426991

3. Romberg integration gives

a) $R_{4,4} = 0.63629437$ **b)** $R_{6,6} = 19.920853$ **c)** $R_{5,5} = 0.78539817$
d) $R_{5,5} = -6.2831855$ **e)** $R_{6,6} = -14.213977$ **f)** $R_{4,4} = 0.47775587$
g) $R_{5,5} = 0.60437561$ **h)** $R_{5,5} = 0.96055026$

4. **a)** $\int_0^1 x^{1/3}\,dx \approx 0.7488276$ **b)** $\int_0^{0.3} f(x)\,dx \approx 0.3024250$

5. Romberg integration gives

a) 62.4373714, 57.2885616, 56.4437507, 56.2630547, and 56.2187727

b) 55.5722917, 56.2014707, 56.2055989, and 56.2040624

c) 58.3626837, 59.0773207, 59.2688746, 59.3175220, 59.3297316, and 59.3327870

d) 58.4220930, 59.4707174, 58.4704791, and 58.4704691

e) Consider the graph of the function.

6. **a)**

$$P_{0,1}(x) = \frac{(x-h^2)N_1(\frac{h}{2})}{\frac{h^2}{4}-h^2} + \frac{(x-\frac{h^2}{4})N_1(h)}{h^2-\frac{h^2}{4}}, \quad \text{so} \quad P_{0,1}(0) = \frac{4N_1(\frac{h}{2})-N_1(h)}{3}$$

Similarly,

$$P_{1,2}(0) = \frac{4N_1(\frac{h}{4})-N_1(\frac{h}{2})}{3}.$$

b)

$$P_{0,2}(x) = \frac{(x-h^4)N_2(\frac{h}{2})}{\frac{h^4}{16}-h^4} + \frac{(x-\frac{h^4}{16})N_2(h)}{h^4-\frac{h^4}{16}}, \quad \text{so} \quad P_{0,2}(0) = \frac{16N_2(\frac{h}{2})-N_2(h)}{15}.$$

EXERCISE SET 4.6 (*Page* 123)

1. The Simpson's rule approximations are

a) $S(1,1.5) = 0.19224530$, $S(1,1.25) = 0.039372434$, $S(1.25,1.5) = 0.15288602$, and the actual value is 0.19225935.

b) $S(0,1) = 0.16240168$, $S(0,0.5) = 0.028861071$, $S(0.5,1) = 0.13186140$, and the actual value is 0.16060279.

c) $S(0,0.35) = -0.17682156$, $S(0,0.175) = -0.087724382$, $S(0.175,0.35) = -0.089095736$, and the actual value is -0.17682002.

d) $S(0,\frac{\pi}{4}) = 0.087995669$, $S(0,\frac{\pi}{8}) = 0.0058315797$, $S(\frac{\pi}{8},\frac{\pi}{4}) = 0.082877624$, and the actual value is 0.088755285.

e) $S(0,\frac{\pi}{4}) = 2.5836964$, $S(0,\frac{\pi}{8}) = 0.033088926$, $S(\frac{\pi}{8},\frac{\pi}{4}) = 2.2568121$, and the actual value is 2.5886286.

f) $S(1,1.6) = -0.73910533$, $S(1,1.3) = -0.26141244$, $S(1.3,1.6) = -0.47305351$, and the actual value is -0.73396917.

g) $S(3,3.5) = 0.63623873$, $S(3,3.25) = 0.32567095$, $S(3.25,3.5) = 0.31054412$, and the actual value is 0.63621334.

h) $S(0,\frac{\pi}{4}) = 0.64326905$, $S(0,\frac{\pi}{8}) = 0.37315002$, $S(\frac{\pi}{8},\frac{\pi}{4}) = 0.26958270$, and the actual value is 0.64269908.

2. Adaptive quadrature gives

a) 0.19226 **b)** 0.16072 **c)** −0.17682 **d)** 0.088709

e) 2.5877 **f)** −0.734466 **g)** 0.636215 **h)** 0.642733

3. Adaptive quadrature gives

a) 108.555281 **b)** −1724.966983 **c)** −15.306308 **d)** −18.945949

4. **a)** Composite Simpson's rule gives −0.21515695 with 434 function evaluations. Adaptive Quadrature gives −0.21515062 with 229 function evaluations.

b) Composite Simpson's rule gives 0.95135137 with 945 function evaluations. Adaptive Quadrature gives 0.95134257 with 217 function evauations.

c) Composite Simpson's rule gives −6.2831813 with 230 function evaluations. Adaptive Quadrature gives −6.2831852 with 109 function evaluations.

d) Composite Simpson's rule gives 5.8696024 with 119 function evaluations. Adaptive Quadrature gives 5.8696044 with 109 function evaluations.

5. Adaptive quadrature gives

$$\int_{0.1}^{2} \sin\frac{1}{x}\,dx \approx 1.1454 \quad \text{and} \quad \int_{0.1}^{2} \cos\frac{1}{x}\,dx \approx 0.67378$$

6. 58.47048 on 593 nodes

7. $\int_1^3 u(t)\,dt \approx -0.13693786$. The graph is shown below on the left.

8. $\int_1^3 u(t)\,dt \approx -0.04328489$. The graph is shown below on the right.

9.

t	$c(t)$	$s(t)$
0.1	0.0999975	0.000523589
0.2	0.199921	0.00418759
0.3	0.299399	0.0141166
0.4	0.397475	0.0333568
0.5	0.492327	0.0647203
0.6	0.581061	0.110498
0.7	0.659650	0.172129
0.8	0.722844	0.249325
0.9	0.764972	0.339747
1.0	0.779880	0.438245

EXERCISE SET 4.7 (*Page* 132)

1. With $n = m = 2$ we have

a) 0.3115733 **b)** 0.2552526 **c)** 16.50864 **d)** 1.476684

2. We have

a) 0.3115733 with $n = m = 1$

b) 0.2552526 with $n = m = 2$

c) 16.50864 with $n = m = 2$

d) no result since it requires $n, m > 400$

3. With $n = 2$ and $m = 4$, $n = 4$ and $m = 2$, and $n = m = 3$ we have

a) 0.5119875, 0.5118533, 0.5118722

b) 1.718857, 1.718220, 1.718385

c) 1.001953, 1.000122, 1.000386

d) 0.7838542, 0.7833659, 0.7834362

e) −1.985611, −1.999182, −1.997353

f) 2.004596, 2.000879, 2.000980

g) 0.3084277, 0.3084562, 0.3084323

h) −22.61612, −19.85408, −20.14117

4. We have

a) 0.51184555 with $n = m = 7$

b) 1.7182827 with $n = m = 10$

c) 1.00000081 with $n = m = 13$

d) 0.78333417 with $n = m = 10$

e) −1.99999913 with $n = m = 22$

f) 2.00000092 with $n = m = 17$

g) 0.30842563 with $n = m = 6$

h) −19.73920977 with $n = m = 72$

5. With $n = m = 2$ we have

a) 0.3115733 **b)** 0.2552446 **c)** 16.50863 **d)** 1.488875

6. We have

a) 0.3115733 with $n = m = 2$ and 4 function evaluations

b) 0.2552519 with $n = m = 3$ and 9 function evaluations

c) 16.508640 with $n = m = 3$ and 9 function evaluations

d) no result, since it requires $n, m > 5$

7. With $n = m = 3$, $n = 3$ and $m = 4$, $n = 4$ and $m = 3$, and $n = m = 4$ we have

a) 0.5118655, 0.5118445, 0.5118655, 0.5118445

b) 1.718163, 1.718302, 1.718139, 1.718277

c) 1.0000000, 1.0000000, 1.000000, 1.000000

d) 0.7833333, 0.7833333, 0.7833333, 0.7833333

e) −1.991878, −2.000124, −1.991878, −2.000124

f) 2.001494, 2.000080, 2.001388, 1.999984

g) 0.3084151, 0.3084145, 0.3084246, 0.3084245

h) −12.74790, −21.21539, −11.83624, −20.30373

8. With $n = m = 5$ we have

a) 0.51184464 **b)** 1.7182816 **c)** 1.0000000 **d)** 0.78333333

e) −1.9999989 **f)** 2.0000001 **g)** 0.30842509 **h)** −19.712428

9. With $n = m = 7$ we have 0.1479099 and with $n = m = 4$ we have 0.1506823.

10. $\iint_R \sqrt{xy + y^2}\, dA \approx 13.15229$

11. a) The area approximation is 1.040253.

b) The area approximation is 1.040252.

12. With $n = m = p = 2$ we have the first listed value. The second value is the exact result.

a) 5.204036, $e(e^{0.5} - 1)(e - 1)^2$ **b)** 0.08429784, $\frac{1}{12}$

c) 0.08641975, $\frac{1}{14}$ **d)** 0.09722222, $\frac{1}{12}$

e) 7.103932, $2 + \frac{1}{2}\pi^2$ **f)** 1.428074, $\frac{1}{2}(e^2 + 1) - e$

13. Gaussian quadrature with $n = m = p = 3$ gives

a) 5.206442 **b)** 0.08333333 **c)** 0.07166667

d) 0.08333333 **e)** 6.928161 **f)** 1.474577

14. With $n = m = p = 4$ we have the first listed value. The second is the result obtained with $n = m = 5$.

a) 5.206447, 5.206447 **b)** 0.08333333, 0.08333333 **c)** 0.07142857, 0.07142857

d) 0.08333333, 0.08333333 **e)** 6.934912, 6.934801 **f)** 1.476207, 1.476246

15. Gaussian quadrature with $n = m = p = 4$ gives 3.052125.

16. The approximation 20.41887 requires 125 functional evaluations.

EXERCISE SET 4.8 (*Page* 138)

1. The Composite Simpson's rule gives

a) 0.5284163 **b)** 4.266654 **c)** 0.4329748 **d)** 0.8802010

2. The Composite Simpson's Rule gives

a) 1.076163 **b)** 0.75 **c)** 1.293701 **d)** 20.07458

3. The Composite Simpson's rule gives

a) 0.4112649 **b)** 0.2440679 **c)** 0.05501681 **d)** 0.2903746

4. The Composite Simpson's Rule gives

a) 1.110722 **b)** 0.5890478

5. The Composite Simpson's rule gives

a) 3.141569 **b)** 0.0 **c)** 1.178071 **d)** 2.221548

6. The escape velocity is approximately 6.9450 miles/sec.

EXERCISE SET 4.9 (*Page* 146)

1. From the forward-backward difference formula we have the following approximations.

a) $f'(0.5) \approx 1.67154$, $f'(0.6) \approx 1.906973$, $f'(0.7) \approx 1.906973$

b) $f'(0.0) \approx 3.5800665$, $f'(0.2) \approx 2.6620555$, $f'(0.4) \approx 2.6620555$

2. a)

x	Actual Error	Error Bound
0.5	0.12350	0.124197
0.6	0.11361	0.119042
0.7	0.10659	0.119042

b)

x	Actual Error	Error Bound
0.0	0.419934	0.459335
0.2	0.478278	0.514726
0.4	0.496762	0.514726

3. For the endpoints of the tables we use the Three-Point Endpoint Formula. The other approximations come from the Three-Point Midpoint Formula.

a) $f'(1.1) \approx 2.557820$, $f'(1.2) \approx 2.448560$, $f'(1.3) \approx 2.352640$, $f'(1.4) \approx 2.270060$

b) $f'(8.1) \approx 3.092050$, $f'(8.3) \approx 3.116150$, $f'(8.5) \approx 3.139975$, $f'(8.7) \approx 3.163525$

c) $f'(2.9) \approx 5.101375$, $f'(3.0) \approx 6.654785$, $f'(3.1) \approx 8.216330$, $f'(3.2) \approx 9.786010$

d) $f'(3.6) \approx -1.45886415$, $f'(3.8) \approx -2.13752785$, $f'(4.0) \approx -2.90294135$, $f'(4.2) \approx -3.75510465$

4. a)

x	Actual Error	Error Bound
1.1	0.011574	0.0153068
1.2	0.0052571	0.00765340
1.3	0.0037137	0.00521092
1.4	0.0068592	0.0104218

b)

x	Actual Error	Error Bound
8.1	0.00018594	0.000040644
8.3	0.00010551	0.000020322
8.5	9.116×10^{-5}	0.000019355
8.7	0.00020197	0.000038709

c)

x	Actual Error	Error Bound
2.9	0.011956	0.0180988
3.0	0.0049251	0.00904938
3.1	0.0004765	0.00493920
3.2	0.0013745	0.00987840

d)

x	Actual Error	Error Bound
3.6	0.0484180	0.0645493
3.8	0.0256291	0.0322747
4.0	0.0323657	0.0408186
4.2	0.0687066	0.0816373

5. The approximations and the formulas used are:

a) $f'(2.1) \approx 3.899344$ from the Five-Point Endpoint Formula, $f'(2.2) \approx 2.876876$ from the Five-Point Endpoint Formula.
$f'(2.3) \approx 2.249704$ from the Five-Point Midpoint Formula, $f'(2.4) \approx 1.837756$ from the Five-Point Midpoint Formula.
$f'(2.5) \approx 1.544210$ from the Five-Point Endpoint Formula, $f'(2.6) \approx 1.355496$ from the Five-Point Endpoint Formula.

b) $f'(-3.0) \approx -5.877358$ from the Five-Point Endpoint Formula, $f'(-2.8) \approx -5.468933$ from the Five-Point Endpoint Formula.
$f'(-2.6) \approx -5.059884$ from the Five-Point Midpoint Formula, $f'(-2.4) \approx -4.650223$ from the Five-Point Midpoint Formula.
$f'(-2.2) \approx -4.239911$ from the Five-Point Endpoint Formula, $f'(-2.0) \approx -3.828853$ from the Five-Point Endpoint Formula.

6. **a)**

x	Actual Error	Error Bound
2.1	0.0242312	0.109271
2.2	0.0105138	0.0386885
2.3	0.0029352	0.0182120
2.4	0.0013262	0.00644808
2.5	0.0138323	0.109271
2.6	0.0064225	0.0386885

b)

x	Actual Error	Error Bound
-3.0	1.55×10^{-5}	6.33×10^{-7}
-2.8	1.32×10^{-5}	6.76×10^{-7}
-2.6	7.95×10^{-7}	1.05×10^{-7}
-2.4	6.79×10^{-7}	1.13×10^{-7}
-2.2	1.28×10^{-5}	6.76×10^{-7}
-2.0	7.96×10^{-6}	6.76×10^{-7}

7. From the forward-backward difference formula we have the following approximations.

a) $f'(0.5) \approx 1.672$, $f'(0.6) \approx 1.907$, $f'(0.7) \approx 1.907$

b) $f'(0.0) \approx 3.580$, $f'(0.2) \approx 2.662$, $f'(0.4) \approx 2.662$

8. **a)**

x	Approximate $f'(x)$	Error
1.1	2.550	0.0194
1.2	2.450	0.0067
1.3	2.355	0.0061
1.4	2.265	0.0119

b)

x	Approximate $f'(x)$	Error
8.1	3.075	0.0168
8.3	3.125	0.0087
8.5	3.150	0.0099
8.7	3.150	0.0133

c)

x	Approximate $f'(x)$	Error
2.9	5.080	0.0333
3.0	6.655	0.0051
3.1	8.220	0.0041
3.2	9.760	0.0246

d)

x	Approximate $f'(x)$	Error
3.6	-1.454	0.0533
3.8	-2.136	0.0241
4.0	-2.902	0.0314
4.2	-3.752	0.0718

9. For the endpoints of the tables we use the Five-Point Endpoint Formula. The other approximations come from the Five-Point Midpoint Formula.

a) $f'(2.1) \approx 3.884$, $f'(2.2) \approx 2.896$, $f'(2.3) \approx 2.249$, $f'(2.4) \approx 1.836$, $f'(2.5) \approx 1.550$, $f'(2.6) \approx 1.348$

b) $f'(-3.0) \approx -5.883$, $f'(-2.8) \approx -5.467$, $f'(-2.6) \approx -5.059$, $f'(-2.4) \approx -4.650$, $f'(-2.2) \approx -4.208$, $f'(-2.0) \approx -3.875$

10. **a)**

	$f'(0.4)$		$f''(0.4)$
$h = 0.6$	-0.8889958	$h = 0.2$	-1.191050
$h = 0.4$	-0.6979043		
$h = 0.2$	-0.5486810		
$h = -0.2$	-0.3104710		
$h = 0.2$	-0.3994578		
$h = 0.2$	-0.4295760		

b)

	$f'(0.6)$		$f''(0.6)$
$h = 0.4$	-1.059153	$h = 0.4$	-1.573943
$h = 0.2$	-0.8471275	$h = 0.2$	-1.492233
$h = -0.2$	-0.5486810		
$h = -0.4$	-0.4295760		
$h = 0.2$	-0.6351018		
$h = -0.2$	-0.6677860		
$h = 0.4$	-0.7443646		
$h = 0.2$	-0.6979043		
$h = 0.2$	-0.6824175		

11. The approximation is -3.10457 with an error bound of 3.98×10^{-2}.

12. With $h = 0.1$ we have 36.641 and with $h = 0.01$ we have 36.5. The actual value is 36.5935.

Chapter 5 Numerical Solution of Initial-Value Problems

EXERCISE SET 5.2 (*Page* 159)

1. Euler's method gives the approximations in the following tables.

a)

i	t_i	w_i	$y(t_i)$	$\|y(t_i) - w_i\|$
0	0.000	0.0000000	0.0000000	0.0000000
1	0.500	0.0000000	0.4160531	0.4160531
2	1.000	1.1204223	3.2678200	2.1473978

b)

i	t_i	w_i	$y(t_i)$	$\|y(t_i) - w_i\|$
0	2.000	1.0000000	1.0000000	0.0000000
1	2.500	2.0000000	1.8333333	0.1666667
2	3.000	2.6250000	2.5000000	0.1250000

c)

i	t_i	w_i	$y(t_i)$	$\|y(t_i) - w_i\|$
0	1.000	2.0000000	2.0000000	0.0000000
1	1.250	2.7500000	2.7789294	0.0289294
2	1.500	3.5500000	3.6081977	0.0581977
3	1.750	4.3916667	4.4793276	0.0876610
4	2.000	5.2690476	5.3862944	0.1172467

d)

i	t_i	w_i	$y(t_i)$	$\|y(t_i) - w_i\|$
0	0.000	1.0000000	1.0000000	0.0000000
1	0.250	1.2500000	1.3291498	0.0791498
2	0.500	1.6398053	1.7304898	0.0906844
3	0.750	2.0242547	2.0414720	0.0172174
4	1.000	2.2364573	2.1179795	0.1184777

2. a)

t	Actual Error	Error bound
0.5	0.416053	11.3938
1.0	2.14740	42.3654

b)

t	Actual Error	Error bound
2.5	0.166667	0.429570
3.0	0.125000	1.59726

c)

t	Actual Error	Error bound
1.25	0.0289294	0.0355032
1.50	0.0581977	0.0810902
1.75	0.0876610	0.139625
2.00	0.117247	0.214785

d)

t	Actual Error	Error bound
0.25	0.0791498	0.289063
0.50	0.0906844	1.24659
0.75	0.0172174	4.41838
1.00	0.118478	14.9250

3. Euler's method gives the approximations in the following tables.

a)

i	t_i	w_i	$y(t_i)$	$\|y(t_i) - w_i\|$
2	1.200	1.0082645	1.0149523	0.0066879
5	1.500	1.0576682	1.0672624	0.0095942
7	1.700	1.1004322	1.1106551	0.0102229
10	2.000	1.1706516	1.1812322	0.0105806

b)

i	t_i	w_i	$y(t_i)$	$\|y(t_i) - w_i\|$
2	1.400	0.4388889	0.4896817	0.0507928
5	2.000	1.4372511	1.6612818	0.2240306
7	2.400	2.4022696	2.8765514	0.4742818
10	3.000	4.5142774	5.8741000	1.3598226

c)

i	t_i	w_i	$y(t_i)$	$\|y(t_i) - w_i\|$
2	0.400	−1.6080000	−1.6200510	0.0120510
5	1.000	−1.1992512	−1.2384058	0.0391546
7	1.400	−1.0797454	−1.1146484	0.0349030
10	2.000	−1.0181518	−1.0359724	0.0178206

d)

i	t_i	w_i	$y(t_i)$	$\lvert y(t_i) - w_i \rvert$
2	0.2	0.1083333	0.1626265	0.0542931
5	0.5	0.2410417	0.2773617	0.0363200
7	0.7	0.4727604	0.5000658	0.0273054
10	1.0	0.9803451	1.0022460	0.0219009

4. The actual errors for the approximations in Exercise 3 are in the following tables.

a)

t	Actual Error
1.2	0.0066879
1.5	0.0095942
1.7	0.0102229
2.0	0.0105806

b)

t	Actual Error
1.4	0.0507928
2.0	0.2240306
2.4	0.4742818
3.0	1.3598226

c)

t	Actual Error
0.4	0.0120510
1.0	0.0391546
1.4	0.0349030
2.0	0.0178206

d)

t	Actual Error
0.2	0.0542931
0.5	0.0363200
0.7	0.0273054
1.0	0.0219009

5. Taylor's method of order two gives the approximations in the following tables.

a)

i	t_i	w_i
1	0.5	0.1250000
2	1.0	2.023239

b)

i	t_i	w_i
1	2.5	1.750000
2	3.0	2.425781

c)

i	t_i	w_i
1	1.25	2.781250
2	1.50	3.612500
3	1.75	4.485417
4	2.00	5.394048

d)

i	t_i	w_i
1	0.25	1.343750
2	0.50	1.772187
3	0.75	2.110676
4	1.00	2.201644

6. a)

t	Taylor Order 2 Approximation	Actual Error
1.2	1.0160294	0.0010770
1.5	1.0686947	0.0014324
1.7	1.1121422	0.0014872
2.0	1.1827427	0.0015105

b)

t	Taylor Order 2 Approximation	Actual Error
1.4	0.4869323	0.0027494
2.0	1.6484426	0.0128392
2.4	2.8461789	0.0303725
3.0	5.7654557	0.1086443

c)

t	Taylor Order 2 Approximation	Actual Error
0.4	−1.6156800	0.0043710
1.0	−1.2362769	0.0021289
1.4	−1.1150276	0.0003792
2.0	−1.0372858	0.0013134

d)

t	Taylor Order 2 Approximation	Actual Error
0.2	0.1702083	0.0075819
0.5	0.2817891	0.0044275
0.7	0.5024176	0.0023518
1.0	1.0030316	0.0007857

7. Taylor's method of order four gives the approximations in the following tables.

a)

i	t_i	w_i	$\|y(t_i) - w_i\|$
3	1.3	1.0298483	3.46×10^{-5}
6	1.6	1.0884694	3.67×10^{-5}
9	1.9	1.1572648	3.64×10^{-5}

b)

i	t_i	w_i	$\|y(t_i) - w_i\|$
3	1.6	0.8126583	9.45×10^{-5}
6	2.2	2.2132495	2.52×10^{-4}
9	2.8	4.6578929	7.72×10^{-4}

c)

i	t_i	w_i	$\lvert y(t_i) - w_i \rvert$
3	0.6	-1.4630012	5.08×10^{-5}
6	1.2	-1.1663295	1.59×10^{-5}
9	1.8	-1.0531846	9.40×10^{-6}

d)

i	t_i	w_i	$\lvert y(t_i) - w_i \rvert$
3	0.3	0.1644651	8.84×10^{-5}
6	0.6	0.3766352	3.95×10^{-5}
9	0.9	0.8137162	1.32×10^{-5}

8. Linear interpolation gives the following results:

a) $1.021957 = y(1.25) \approx 1.014978$, $1.164390 = y(1.93) \approx 1.153902$

b) $1.924962 = y(2.1) \approx 1.660756$, $4.394170 = y(2.75) \approx 3.526160$

c) $-1.138277 = y(1.3) \approx -1.103618$, $-1.041267 = y(1.93) \approx -1.028184$

d) $0.3140018 = y(0.54) \approx 0.2828333$, $0.8866318 = y(0.94) \approx 0.8665521$

9. Linear interpolation gives the following results:

a) $1.021957 = y(1.25) \approx 1.023555$; $1.164390 = y(1.93) \approx 1.165937$

b) $1.924962 = y(2.1) \approx 1.921037$; $4.394170 = y(2.75) \approx 4.349162$

c) $-1.114648 = y(1.4) \approx -1.115028$; $-1.041267 = y(1.93) \approx -1.043301$

d) $0.3140018 = y(0.54) \approx 0.3210207$; $0.9108858 = y(0.94) \approx 0.8901230$

10. Cubic Hermite interpolation gives the following results:

a) $1.021957 = y(1.25) \approx 1.023127$; $1.164390 = y(1.93) \approx 1.165896$

b) $1.924962 = y(2.1) \approx 1.908843$; $4.394170 = y(2.75) \approx 4.329816$

c) $-1.138277 = y(1.3) \approx -1.138168$; $-1.041267 = y(1.93) \approx -1.042605$

d) $0.3140018 = y(0.54) \approx 0.3178578$; $0.9108858 = y(0.94) \approx 0.8876066$

11.

j	t_j	w_j
20	2	0.702938
40	4	-0.0457793
60	6	0.294870
80	8	0.341673
100	10	0.139432

12. **a)** Since $p(t) = x_n(t)/x(t)$ we have

$$\begin{aligned}\frac{dp(t)}{dt} &= \frac{-x_n(t)x'(t)}{[x(t)]^2} + \frac{x_n'(t)}{x_n(t)} \\ &= \frac{-(b-d)x_n(t)}{x(t)} + \frac{(b-d)x_n(t) + rb(x(t) - x_n(t))}{x(t)} \\ &= rb\left(1 - \frac{x_n(t)}{x(t)}\right) = rb(1 - p(t))\end{aligned}$$

b) $w_{50} = 0.10430 \approx p(50)$

c) Since $p(t) = 1 - 0.99e^{-0.002t}$, $p(50) = 0.10421$.

13. **a)** Taylors method of order 2 gives the table on the left. The table on the right comes from Taylors method of order 4.

i	t_i	w_i
2	0.2	5.86595
5	0.5	2.82145
7	0.7	0.84926
10	1.0	−2.08606

i	t_i	w_i
2	0.2	5.86433
5	0.5	2.81789
7	0.7	0.84455
10	1.0	−2.09015

b) The projectile reaches its maximum height after 0.8 seconds

EXERCISE SET 5.3 (*Page* 166)

1. **a)**

t	Modified Euler	$y(t)$	Error
0.5	0.5602111	0.2836165	0.2765946
1.0	5.3014898	3.2190993	2.0823905

b)

t	Modified Euler	$y(t)$	Error
2.5	1.8125000	1.8333333	0.0208333
3.0	2.4815531	2.5000000	0.0184469

c)

t	Modified Euler	$y(t)$	Error
1.25	2.7750000	2.7789294	0.0039294
1.50	3.6008333	3.6081977	0.0073643
1.75	4.4688294	4.4793276	0.0104983
2.00	5.3728586	5.3862944	0.0134358

d)

t	Modified Euler	$y(t)$	Error
0.25	1.3199027	1.3291498	0.0092471
0.50	1.7070300	1.7304898	0.0234598
0.75	2.0053560	2.0414720	0.0361161
1.00	2.0770789	2.1179795	0.0409006

2. a)

t	Heun	$y(t)$	Error
0.50	0.3397852	0.2836165	0.0561687
1.00	3.6968164	3.2190993	0.4777171

b)

t	Heun	$y(t)$	Error
2.50	1.7916667	1.8333333	0.0416667
3.00	2.4641747	2.5000000	0.0358253

c)

t	Heun	$y(t)$	Error
1.25	2.7767857	2.7789294	0.0021437
1.50	3.6042017	3.6081977	0.0039960
1.75	4.4736520	4.4793276	0.0056757
2.00	5.3790494	5.3862944	0.0072449

d)

t	Heun	$y(t)$	Error
0.25	1.3295717	1.3291498	0.0004219
0.50	1.7310350	1.7304898	0.0005452
0.75	2.0417476	2.0414720	0.0002756
1.00	2.1176975	2.1179795	0.0002820

3. a)

t	Midpoint	$y(t)$	Error
0.5	0.2646250	0.2836165	0.0189915
1.0	3.1300023	3.2190993	0.0890970

b)

t	Midpoint	$y(t)$	Error
2.5	1.7812500	1.8333333	0.0520833
3.0	2.4550638	2.5000000	0.0449362

c)

t	Midpoint	$y(t)$	Error
1.25	2.7777778	2.7789294	0.0011517
1.50	3.6060606	3.6081977	0.0021371
1.75	4.4763015	4.4793276	0.0030262
2.00	5.3824398	5.3862944	0.0038546

d)

t	Midpoint	$y(t)$	Error
0.25	1.3337962	1.3291498	0.0046464
0.50	1.7422854	1.7304898	0.0117956
0.75	2.0596374	2.0414720	0.0181654
1.00	2.1385560	2.1179795	0.0205764

4. a)

t	Modified Euler	$y(t)$	Error
1.2	1.0147137	1.0149523	0.0002386
1.5	1.0669093	1.0672624	0.0003530
1.7	1.1102751	1.1106551	0.0003800
2.0	1.1808345	1.1812322	0.0003977

b)

t	Modified Euler	$y(t)$	Error
1.4	0.4850495	0.4896817	0.0046322
2.0	1.6384229	1.6612818	0.0228589
2.4	2.8250651	2.8765514	0.0514863
3.0	5.7075699	5.8741000	0.1665301

c)

t	Modified Euler	$y(t)$	Error
0.4	−1.6229206	−1.6200510	0.0028696
1.0	−1.2442903	−1.2384058	0.0058845
1.4	−1.1200763	−1.1146484	0.0054280
2.0	−1.0391938	−1.0359724	0.0032214

d)

t	Modified Euler	$y(t)$	Error
0.2	0.1742708	0.1626265	0.0116444
0.5	0.2878200	0.2773617	0.0104584
0.7	0.5088359	0.5000658	0.0087702
1.0	1.0096377	1.0022460	0.0073917

5. Linear interpolation gives the following results:

a) $1.0221167 \approx y(1.25) = 1.0219569,\ 1.1640347 \approx y(1.93) = 1.1643901$

b) $1.9086500 \approx y(2.1) = 1.9249616,\ 4.3105913 \approx y(2.75) = 4.3941697$

c) $-1.1461434 \approx y(1.3) = -1.1382768,\ -1.0454854 \approx y(1.93) = -1.0412665$

d) $0.3271470 \approx y(0.54) = 0.3140018,\ 0.8967073 \approx y(0.94) = 0.8866318$

6. **a)**

t	Heun	$y(t)$	Error
1.2	1.0151123	1.0149523	0.0001600
1.5	1.0674528	1.0672624	0.0001904
1.7	1.1108444	1.1106551	0.0001894
2.0	1.1814172	1.1812322	0.0001850

b)

t	Heun	$y(t)$	Error
1.4	0.4858314	0.4896817	0.0038502
2.0	1.6421387	1.6612818	0.0191431
2.4	2.8327728	2.8765514	0.0437787
3.0	5.7286247	5.8741000	0.1454753

c)

t	Heun	$y(t)$	Error
0.4	−1.6205037	−1.6200510	0.0004527
1.0	−1.2415866	−1.2384058	0.0031807
1.4	−1.1183618	−1.1146484	0.0037134
2.0	−1.0385425	−1.0359724	0.0025701

d)

t	Heun	$y(t)$	Error
0.2	0.1729167	0.1626265	0.0102902
0.5	0.2858097	0.2773617	0.0084481
0.7	0.5066965	0.5000658	0.0066307
1.0	1.0074357	1.0022460	0.0051897

7. Linear interpolation gives the following results:

a) $1.0225530 \approx y(1.25) = 1.0219569,\ 1.1646155 \approx y(1.93) = 1.1643901$

b) $1.9132167 \approx y(2.1) = 1.9249616,\ 4.3246152 \approx y(2.75) = 4.3941697$

c) $-1.1441775 \approx y(1.3) = -1.1382768,\ -1.0447403 \approx y(1.93) = -1.0412665$

d) $0.3251049 \approx y(0.54) = 0.3140018,\ 0.8945125 \approx y(0.94) = 0.8866318$

8. **a)**

t	Midpoint	$y(t)$	Error
1.2	1.0153257	1.0149523	0.0003734
1.5	1.0677427	1.0672624	0.0004804
1.7	1.1111478	1.1106551	0.0004928
2.0	1.1817275	1.1812322	0.0004952

b)

t	Midpoint	$y(t)$	Error
1.4	0.4861770	0.4896817	0.0035046
2.0	1.6438889	1.6612818	0.0173928
2.4	2.8364357	2.8765514	0.0401157
3.0	5.7386475	5.8741000	0.1354525

c)

t	Midpoint	$y(t)$	Error
0.4	−1.6192966	−1.6200510	0.0007545
1.0	−1.2402470	−1.2384058	0.0018411
1.4	−1.1175165	−1.1146484	0.0028681
2.0	−1.0382227	−1.0359724	0.0022503

d)

t	Midpoint	$y(t)$	Error
0.2	0.1722396	0.1626265	0.0096131
0.5	0.2848046	0.2773617	0.0074429
0.7	0.5056268	0.5000658	0.0055610
1.0	1.0063347	1.0022460	0.0040887

9. Linear interpolation gives the following results:

a) $1.0227863 \approx y(1.25) = 1.0219569,\ 1.1649247 \approx y(1.93) = 1.1643901$

b) $1.9153749 \approx y(2.1) = 1.9249616,\ 4.3312939 \approx y(2.75) = 4.3941697$

c) $-1.1432070 \approx y(1.3) = -1.1382768,\ -1.0443743 \approx y(1.93) = -1.0412665$

d) $0.3240839 \approx y(0.54) = 0.3140018,\ 0.8934152 \approx y(0.94) = 0.8866318$

10. a)

t	Runge-Kutta	$y(t)$	Error
0.5	0.2969975	0.2836165	1.33809×10^{-2}
1.0	3.3143118	3.2190993	9.52125×10^{-2}

b)

t	Runge-Kutta	$y(t)$	Error
2.5	1.8333234	1.8333333	9.97260×10^{-6}
3.0	2.4999712	2.5000000	2.88066×10^{-5}

c)

t	Runge-Kutta	$y(t)$	Error
1.3	2.7789095	2.7789294	1.99741×10^{-5}
1.5	3.6081647	3.6081977	3.29341×10^{-5}
1.8	4.4792846	4.4793276	4.30257×10^{-5}
2.0	5.3862426	5.3862944	5.17723×10^{-5}

d)

t	Runge-Kutta	$y(t)$	Error
0.3	1.3291650	1.3291498	1.52339×10^{-5}
0.5	1.7305336	1.7304898	4.38078×10^{-5}
0.8	2.0415436	2.0414720	7.15432×10^{-5}
1.0	2.1180636	2.1179795	8.40544×10^{-5}

11. a)

t	Runge-Kutta	$y(t)$	Error
1.2	1.0149520	1.0149523	3.10759×10^{-7}
1.5	1.0672620	1.0672624	3.66382×10^{-7}
1.7	1.1106547	1.1106551	3.66990×10^{-7}
2.0	1.1812319	1.1812322	3.62576×10^{-7}

b)

t	Runge-Kutta	$y(t)$	Error
1.4	0.4896842	0.4896817	2.50269×10^{-6}
2.0	1.6612651	1.6612818	1.66403×10^{-5}
2.4	2.8764941	2.8765514	5.73055×10^{-5}
3.0	5.8738386	5.8741000	2.61408×10^{-4}

c)

t	Runge-Kutta	$y(t)$	Error
0.4	−1.6200576	−1.6200510	6.60106×10^{-6}
1.0	−1.2384307	−1.2384058	2.48941×10^{-5}
1.4	−1.1146769	−1.1146484	2.85862×10^{-5}
2.0	−1.0359922	−1.0359724	1.98032×10^{-5}

d)

t	Runge-Kutta	$y(t)$	Error
0.2	0.1627655	0.1626265	1.38977×10^{-4}
0.5	0.2774767	0.2773617	1.14994×10^{-4}
0.7	0.5001579	0.5000658	9.21539×10^{-5}
1.0	1.0023207	1.0022460	7.46867×10^{-5}

12. Cubic Hermite interpolation gives the following results:

a) $1.021957 = y(1.25) \approx 1.021955$, $1.164390 = y(1.93) \approx 1.164390$

b) $1.924962 = y(2.10) \approx 1.924921$, $4.394170 = y(2.75) \approx 4.393994$

c) $-1.138277 = y(1.3) \approx -1.138304$, $-1.041267 = y(1.93) \approx -1.041286$

d) $0.3140018 = y(0.54) \approx 0.3141058$, $0.8866318 = y(0.94) \approx 0.8867066$

13. **a)** 6.531327 ft **b)** 25 min

14. In 0.2 seconds we have approximately 2099 units of KOH.

EXERCISE SET 5.4 (*Page* 174)

1. The Adams-Bashforth methods give the results in the following tables.

a)

t	2 step	3 step	4 step	5 step	$y(t)$
0.2	0.0268128	0.0268128	0.0268128	0.0268128	0.0268128
0.4	0.1200522	0.1507778	0.1507778	0.1507778	0.1507778
0.6	0.4153551	0.4613866	0.4960196	0.4960196	0.4960196
0.8	1.1462844	1.2512447	1.2961260	1.3308570	1.3308570
1.0	2.8241683	3.0360680	3.1461400	3.1854002	3.2190993

b)

t	2 step	3 step	4 step	5 step	$y(t)$
2.2	1.3666667	1.3666667	1.3666667	1.3666667	1.3666667
2.4	1.6750000	1.6857143	1.6857143	1.6857143	1.6857143
2.6	1.9632431	1.9794407	1.9750000	1.9750000	1.9750000
2.8	2.2323184	2.2488759	2.2423065	2.2444444	2.2444444
3.0	2.4884512	2.5051340	2.4980306	2.5011406	2.5000000

c)

t	2 step	3 step	4 step	5 step	$y(t)$
1.2	2.6187859	2.6187859	2.6187859	2.6187859	2.6187859
1.4	3.2734823	3.2710611	3.2710611	3.2710611	3.2710611
1.6	3.9567107	3.9514231	3.9520058	3.9520058	3.9520058
1.8	4.6647738	4.6569191	4.6582078	4.6580160	4.6580160
2.0	5.3949416	5.3848058	5.3866452	5.3862177	5.3862944

d)

t	2 step	3 step	4 step	5 step	$y(t)$
0.2	1.2529306	1.2529306	1.2529306	1.2529306	1.2529306
0.4	1.5986417	1.5712255	1.5712255	1.5712255	1.5712255
0.6	1.9386951	1.8827238	1.8750869	1.8750869	1.8750869
0.8	2.1766821	2.0844122	2.0698063	2.0789180	2.0789180
1.0	2.2369407	2.1115540	2.0998117	2.1180642	2.1179795

2. (1a)

t	2 step	3 step	4 step	$y(t)$
0.2	0.0268128	0.0268128	0.0268128	0.0268128
0.4	0.1533627	0.1507778	0.1507778	0.1507778
0.6	0.5030068	0.4979042	0.4960196	0.4960196
0.8	1.3463142	1.3357923	1.3322919	1.3308570
1.0	3.2512866	3.2298092	3.2227484	3.2190993

(1c)

t	2 step	3 step	4 step	$y(t)$
1.2	2.6187859	2.6187859	2.6187859	2.6187859
1.4	3.2711394	3.2710611	3.2710611	3.2710611
1.6	3.9521454	3.9519886	3.9520058	3.9520058
1.8	4.6582064	4.6579866	4.6580211	4.6580160
2.0	5.3865293	5.3862558	5.3863027	5.3862944

(1d)

t	2 step	3 step	4 step	$y(t)$
0.2	1.2529306	1.2529306	1.2529306	1.2529306
0.4	1.5700866	1.5712255	1.5712255	1.5712255
0.6	1.8738414	1.8757546	1.8750869	1.8750869
0.8	2.0787117	2.0803067	2.0789471	2.0789180
1.0	2.1196912	2.1199024	2.1178679	2.1179795

3. The Adams-Bashforth methods give the results in the following tables.

a)

t	2 step	3 step	4 step	5 step	$y(t)$
1.2	1.0161982	1.0149520	1.0149520	1.0149520	1.0149523
1.5	1.0697141	1.0664788	1.0675362	1.0671695	1.0672624
1.7	1.1133294	1.1097691	1.1109994	1.1105036	1.1106551
2.0	1.1840272	1.1803057	1.1815967	1.1810689	1.1812322

b)

t	2 step	3 step	4 step	5 step	$y(t)$
1.4	0.4867550	0.4896842	0.4896842	0.4896842	0.4896817
2.0	1.6377944	1.6584313	1.6603060	1.6613179	1.6612818
2.4	2.8163947	2.8667672	2.8735320	2.8762776	2.8765514
3.0	5.6491203	5.8268008	5.8589944	5.8706101	5.8741000

c)

t	2 step	3 step	4 step	5 step	$y(t)$
0.5	−1.5357010	−1.5381988	−1.5379372	−1.5378676	−1.5378828
1.0	−1.2374093	−1.2389605	−1.2383734	−1.2383693	−1.2384058
1.5	−1.0952910	−1.0950952	−1.0947925	−1.0948481	−1.0948517
2.0	−1.0366643	−1.0359996	−1.0359497	−1.0359760	−1.0359724

d)

t	2 step	3 step	4 step	5 step	$y(t)$
0.2	0.1739041	0.1627655	0.1627655	0.1627655	0.1626265
0.5	0.2846336	0.2732179	0.2780929	0.2769031	0.2773617
0.7	0.5042285	0.4972078	0.4998405	0.4988777	0.5000658
1.0	1.0037415	1.0020894	1.0064121	1.0073348	1.0022460

4. a)

t	w	$y(t)$	Error
0.0	0.0000000	0.0000000	0
0.2	0.0269059	0.0268128	0.0000931
0.4	0.1510468	0.1507778	0.0002690
0.6	0.4966479	0.4960196	0.0006283
0.8	1.3408657	1.3308570	0.0100087
1.0	3.2450881	3.2190993	0.0259888

b)

t	w	$y(t)$	Error
2.2	1.3666610	1.3666667	0.0000057
2.4	1.6857079	1.6857143	0.0000064
2.6	1.9749941	1.9750000	0.0000059
2.8	2.2446995	2.2444444	0.0002550
3.0	2.5003083	2.5000000	0.0003083

c)

t	w	$y(t)$	Error
1.2	2.6187787	2.6187859	0.0000072
1.4	3.2710491	3.2710611	0.0000120
1.6	3.9519900	3.9520058	0.0000159
1.8	4.6579968	4.6580160	0.0000191
2.0	5.3862715	5.3862944	0.0000228

d)

t	w	$y(t)$	Error
0.2	1.2529350	1.2529306	0.0000044
0.4	1.5712383	1.5712255	0.0000129
0.6	1.8751097	1.8750869	0.0000228
0.8	2.0796618	2.0789180	0.0007437
1.0	2.1192575	2.1179795	0.0012779

5. The Adams Fourth-Order Predictor-Corrector method gives the results in the following tables.

a)

t	w	$y(t)$
1.2	1.0149520	1.0149523
1.5	1.0672462	1.0672624
1.7	1.1106352	1.1106551
2.0	1.1812112	1.1812322

b)

t	w	$y(t)$
1.4	0.4896842	0.4896817
2.0	1.6612586	1.6612818
2.4	2.8765082	2.8765514
3.0	5.8739518	5.8741000

c)

t	w	$y(t)$
0.5	−1.5378788	−1.5378828
1.0	−1.2384134	−1.2384058
1.5	−1.0948609	−1.0948517
2.0	−1.0359757	−1.0359724

d)

t	w	$y(t)$
0.2	0.1627655	0.1626265
0.5	0.2769896	0.2773617
0.7	0.4998012	0.5000658
1.0	1.0021372	1.0022460

6. **a)** With $h = 0.01$, the three-step Adams-Moulton method gives the values in the following table.

i	t_i	w_i
10	0.1	1.317218
20	0.2	1.784511

b) Newton's method will reduce the number of iterations per step from 4 to 3, using the stopping criterion

$$|w_i^{(k)} - w_i^{(k-1)}| \le 10^{-6}.$$

7. Milne-Simpson's Predictor-Corrector method gives the results in the following tables.

a)

i	t_i	w_i	$y(t)$
2	1.2	1.01495200	1.01495231
5	1.5	1.06725997	1.06726235
7	1.7	1.11065221	1.11065505
10	2.0	1.18122584	1.18123222

b)

i	t_i	w_i	$y(t)$
2	1.4	0.48968417	0.48968166
5	2.0	1.66126150	1.66128176
7	2.4	2.87648763	2.87655142
10	3.0	5.87375555	5.87409998

c)

i	t_i	w_i	$y(t)$
5	0.5	−1.53788255	−1.53788284
10	1.0	−1.23840789	−1.23840584
15	1.5	−1.09485532	−1.09485175
20	2.0	−1.03597247	−1.03597242

d)

i	t_i	w_i	$y(t)$
2	0.2	0.16276546	0.16262648
5	0.5	0.27741080	0.27736167
7	0.7	0.50008713	0.50006579
10	1.0	1.00215439	1.00224598

8. $h = 0.1$

i	$t(i)$	$w(i)$	$y(t)$	Error
5	0.50000000	0.22323512	0.22313016	1.0495660×10^{-4}
10	1.00000000	0.01759886	0.01831564	7.1677443×10^{-4}
15	1.50000000	0.00143997	0.00150344	6.3470763×10^{-5}
20	2.00000000	0.00014508	0.00012341	2.1674393×10^{-5}

$h = 0.05$

i	$t(i)$	$w(i)$	$y(t)$	Error
5	0.25000000	0.77880158	0.77880078	7.9795149×10^{-7}
10	0.50000000	0.22305989	0.22313016	7.0268844×10^{-5}
15	0.75000000	0.06394268	0.06392786	1.4823625×10^{-5}
20	1.00000000	0.01826938	0.01831564	4.6259158×10^{-5}
25	1.25000000	0.00528604	0.00524752	3.8519648×10^{-5}
30	1.50000000	0.00145680	0.00150344	4.6636499×10^{-5}
35	1.75000000	0.00048022	0.00043074	4.9475404×10^{-5}
40	2.00000000	0.00006872	0.00012341	5.4685031×10^{-5}

With the smaller value for h, the actual error seems to increase. We would expect the error to decrease.

EXERCISE SET 5.5 (*Page* 178)

1. The Extrapolation method gives the results in the following tables.

a)

t_i	w_i	h_i	$y(t_i)$
0.25	0.0454263	0.25	0.0454312
0.50	0.2835997	0.25	0.2836165
0.75	1.0525685	0.25	1.0525761
1.00	3.2190944	0.25	3.2190993

b)

t_i	w_i	h_i	$y(t_i)$
2.25	1.4499819	0.25	1.4500000
2.50	1.8333386	0.25	1.8333333
2.75	2.1785826	0.25	2.1785714
3.00	2.5000119	0.25	2.5000000

c)

t_i	w_i	h_i	$y(t_i)$
1.25	2.7789152	0.25	2.7789294
1.50	3.6081723	0.25	3.6081977
1.75	4.4792929	0.25	4.4793276
2.00	5.3862512	0.25	5.3862944

d)

t_i	w_i	h_i	$y(t_i)$
0.25	1.3291498	0.25	1.3291498
0.50	1.7304898	0.25	1.7304898
0.75	2.0414720	0.25	2.0414720
1.00	2.1179795	0.25	2.1179795

2. The Extrapolation method gives the results in the following tables.

a)

t_i	w_i	h_i	$y(t_i)$
1.50	1.0672624	0.50	1.0672624
2.00	1.1812322	0.50	1.1812322
2.50	1.3046035	0.50	1.3046037
3.00	1.4295159	0.50	1.4295161
3.50	1.5536475	0.50	1.5536477
4.00	1.6762389	0.50	1.6762391

b)

t_i	w_i	h_i	$y(t_i)$
1.50	0.6438753	0.50	0.6438753
2.00	1.6612817	0.50	1.6612818
2.50	3.2580152	0.50	3.2580154
3.00	5.8740997	0.50	5.8741000

c)

t_i	w_i	h_i	$y(t_i)$
0.50	−1.5378828	0.50	−1.5378828
1.00	−1.2384058	0.50	−1.2384058
1.50	−1.0948517	0.50	−1.0948517
2.00	−1.0359724	0.50	−1.0359724
2.50	−1.0133857	0.50	−1.0133857
3.00	−1.0049452	0.50	−1.0049452

d)

t_i	w_i	h_i	$y(t_i)$
0.50	0.2987518	0.50	0.2987518
1.00	0.2166264	0.50	0.2166264
1.50	0.1245857	0.50	0.1245856
2.00	0.0543455	0.50	0.0543455

3. a)

t	w	$y(t)$
1.972454	−0.269257	−0.269257
2.172454	−0.398491	−0.398491
2.372454	−0.346753	−0.346753
2.572454	−0.185816	−0.185816
2.772454	−0.086328	−0.086328
2.972454	−0.170751	−0.170751
3.141593	−0.323039	−0.323039

b)

t	w	$y(t)$
1.770796	1.603594	1.603594
1.970796	2.876235	2.876235
2.170796	4.753496	4.753496
2.356194	7.540468	7.540468

c)

t	w	$y(t)$
2.918282	2.043602	2.043602
3.118282	2.093943	2.093943
3.318282	2.148044	2.148043
3.518282	2.204142	2.204141
3.718282	2.261173	2.261173
3.918282	2.318486	2.318486
4.118282	2.375674	2.375674
4.318282	2.432486	2.432486
4.518282	2.488767	2.488767
4.718282	2.544424	2.544424
4.918282	2.599404	2.599404
5.118282	2.653680	2.653680
5.318282	2.707245	2.707245
5.436564	2.738589	2.738589

d)

t	w	$y(t)$
0.2	0.219998	0.219997
0.4	0.479830	0.479830
0.6	0.778081	0.778081
0.8	1.109507	1.109507
1.0	1.462117	1.462117
1.2	1.816909	1.816909
1.4	2.153066	2.153066
1.6	2.456485	2.456485
1.8	2.724624	2.724624
2.0	2.964028	2.964028
2.2	3.184310	3.184310
2.4	3.393718	3.393718
2.6	3.597684	3.597684
2.8	3.799213	3.799213
3.0	3.999753	3.999753

4. The Extrapolation method gives the results in the following tables.

a)

t_i	w_i	h_i	$y(t_i)$
1.200	1.4505993	0.200	1.4506019
1.400	1.2436158	0.200	1.2436200
1.600	1.3544829	0.200	1.3544774
1.800	2.8635071	0.200	2.8634805
1.850	5.5246259	0.050	5.5245216
1.875	11.8864275	0.025	11.8864757

$HMIN = 0.02$ exceeded

b)

t_i	w_i	h_i	$y(t_i)$
0.20	0.3105581	0.2	0.3105590
0.40	0.6678406	0.2	0.6678413
0.50	0.8544209	0.1	0.8544225

c)

t_i	w_i	h_i	$y(t_i)$
0.2	1.9404008	0.2	1.9403997
0.4	1.7663863	0.2	1.7663840
0.5	1.6405703	0.1	1.6405667

d)

t_i	w_i	h_i	$y(t_i)$
0.2	0.1626265	0.2	0.1626265
0.4	0.2051115	0.2	0.2051118
0.6	0.3765955	0.2	0.3765957
0.8	0.6461050	0.2	0.6461052
1.0	1.0022458	0.2	1.0022460

5. $y(5) \approx 56,751$.

EXERCISE SET 5.6 (*Page* 186)

1. The Runge-Kutta-Fehlberg method gives the results in the following tables.

a)

t	w	h	$y(t)$
0.2093900	0.0298184	0.2093900	0.0298337
0.3832972	0.1343260	0.1739072	0.1343488
0.5610469	0.4016438	0.1777496	0.4016860
0.7106840	0.8708372	0.1496371	0.8708882
0.8387744	1.5894061	0.1280905	1.5894600
0.9513263	2.6140226	0.1125519	2.6140771
1.0000000	3.2190497	0.0486737	3.2190993

b)

t	w	h	$y(t)$
2.25	1.4499988	0.25	1.4500000
2.50	1.8333332	0.25	1.8333333
2.75	2.1785718	0.25	2.1785714
3.00	2.5000005	0.25	2.5000000

c)

t	w	h	$y(t)$
1.25	2.7789299	0.25	2.7789294
1.50	3.6081985	0.25	3.6081977
1.75	4.4793288	0.25	4.4793276
2.00	5.3862958	0.25	5.3862944

d)

t	w	h	$y(t)$
0.25	1.3291478	0.25	1.3291498
0.50	1.7304857	0.25	1.7304898
0.75	2.0414669	0.25	2.0414720
1.00	2.1179750	0.25	2.1179795

2. The Runge-Kutta-Fehlberg method gives the results in the following tables.

a)

t_i	w_i	h	$y(t)$
1.1101946	1.0051237	0.1101946	1.0051237
1.3572694	1.0396749	0.1381381	1.0396749
1.7470584	1.1213948	0.2180472	1.1213947
2.3994350	1.2795396	0.3707934	1.2795395
3.3985147	1.5285639	0.5000000	1.5285638
4.0000000	1.6762393	0.1014853	1.6762391

b)

t_i	w_i	h	$y(t)$
1.1450265	0.1560235	0.1450265	0.1560234
1.5482238	0.7234123	0.1256486	0.7234119
1.8847226	1.3851234	0.1073571	1.3851226
2.1846024	2.1673514	0.0965027	2.1673499
2.3678493	2.7614423	0.0900249	2.7614402
2.6193834	3.7649220	0.0808140	3.7649189
2.8443728	4.9055656	0.0721522	4.9055611
3.0000000	5.8741059	0.0195070	5.8741000

c)

t_i	w_i	h	$y(t)$
0.1633541	−1.8380836	0.1633541	−1.8380836
0.6319516	−1.4406054	0.1255259	−1.4406054
0.9823055	−1.2459376	0.1088025	−1.2459378
1.2980161	−1.1387882	0.1049836	−1.1387884
1.7370979	−1.0601106	0.1141628	−1.0601108
2.1074733	−1.0291158	0.1288897	−1.0291161
2.5379042	−1.0124143	0.1518525	−1.0124145
3.0000000	−1.0049450	0.1264618	−1.0049452

d)

t_i	w_i	h	$y(t)$
0.3986051	0.3108201	0.3986051	0.3108199
0.6837259	0.2720572	0.2851208	0.2720570
0.9703970	0.2221189	0.2866710	0.2221186
1.2630819	0.1671979	0.2926849	0.1671973
1.5672905	0.1133085	0.3042087	0.1133082
1.7333071	0.0876899	0.1660166	0.0876897
1.9097698	0.0644823	0.1764627	0.0644824
2.0000000	0.0543454	0.0902302	0.0543455

3. The Runge-Kutta-Fehlberg method gives the results in the following tables.

a)

t	w	h	$y(t)$
1.070211	1.755041	0.0702106	1.755040
1.151431	1.545701	0.0812206	1.545701
1.233265	1.396559	0.0818337	1.396557
1.323237	1.290561	0.0899720	1.290559
1.432414	1.236108	0.1091769	1.236106
1.527855	1.264192	0.0954415	1.264189
1.612733	1.379554	0.0848773	1.379548
1.661500	1.513496	0.0487677	1.513488
1.690343	1.632540	0.0288423	1.632530
1.716154	1.777972	0.0258112	1.777960
1.736435	1.929852	0.0202807	1.929837

b)

t	w	h	$y(t)$
0.2	0.310560	0.2	0.310560
0.4	0.667845	0.2	0.667841
0.5	0.854426	0.1	0.854423

c)

t	w	h	$y(t)$
0.2	1.940400	0.2	1.940400
0.4	1.766384	0.2	1.766384
0.5	1.640567	0.1	1.640567

d)

t	w	h	$y(t)$
0.0384750	0.276479	0.038475	0.276480
0.0795096	0.230310	0.041035	0.230310
0.3095255	0.166723	0.049442	0.166723
0.4676371	0.250851	0.054245	0.250852
0.5232524	0.298151	0.055615	0.298152
0.6379535	0.420711	0.057873	0.420712
0.7561692	0.579392	0.059467	0.579393
0.8767031	0.772768	0.060493	0.772769
1.0000000	1.002245	0.001344	1.002246

4. The following tables list representative results from the Adams Variable Step-Size Predictor-Corrector method, including the values at each step size change.

a)

t	w	h	$y(t)$
0.0427560	0.0009689	0.0427560	0.0009689
0.2249146	0.0352944	0.0538908	0.0352936
0.6456078	0.6292294	0.0434579	0.6292053
0.8552122	1.7119813	0.0357729	1.7119543
0.9987279	3.2018104	0.0004240	3.2017819
1.0000000	3.2191278	0.0004240	3.2190993

b)

t	w	h	$y(t)$
2.0625000	1.1213235	0.0625000	1.1213235
2.5311096	1.8779922	0.0936096	1.8779885
2.9993683	2.4992157	0.0002106	2.4992103
3.0000000	2.5000054	0.0002106	2.5000000

c)

t	w	h	$y(t)$
1.0625000	2.1894136	0.0625000	2.1894137
1.4002564	3.2719156	0.1502564	3.2719166
1.8882692	4.9768337	0.0372436	4.9768368
2.0000000	5.3862911	0.0372436	5.3862944

d)

t	w	h	$y(t)$
0.0625	1.0681796	0.0625	1.0681796
0.2500	1.3291512	0.0625	1.3291498
0.5000	1.7304990	0.0625	1.7304898
0.7500	2.0414898	0.0625	2.0414720
1.0000	2.1180021	0.0625	2.1179795

5. The following tables list representative results from the Adams Variable Step-Size Predictor-Corrector method, including the values at each step size change.

a)

t	w	h	σ	$y(t)$
1.0208633	1.0002103	0.0208633	8.758×10^{-7}	1.0002103
1.4067597	1.0488117	0.0312203	2.704×10^{-7}	1.0488118
1.7672125	1.1260408	0.0482499	2.478×10^{-7}	1.1260409
2.8465794	1.3912095	0.1110703	1.923×10^{-7}	1.3912097
3.6347511	1.5868527	0.1217496	4.863×10^{-8}	1.5868529
4.0000000	1.6762389	0.1217496	4.863×10^{-8}	1.6762391

b)

t	w	h	σ	$y(t)$
1.0370392	0.0377335	0.0370392	5.885×10^{-7}	0.0377335
1.5555881	0.7358664	0.0370392	1.400×10^{-7}	0.7358663
2.0000586	1.6614297	0.0370392	2.385×10^{-7}	1.6614294
2.6236064	3.7839775	0.0309205	6.569×10^{-7}	3.7839765
2.8658904	5.0296482	0.0258405	6.460×10^{-7}	5.0296466
2.9950929	5.8408682	0.0258405	7.607×10^{-7}	5.8408661

c)

t	w	h	σ	$y(t)$
0.0337080	−1.9663047	0.0337080	5.243×10^{-7}	−1.9663047
0.4907365	−1.5451991	0.0525323	9.494×10^{-7}	−1.5451992
0.6902430	−1.4018619	0.0419096	6.326×10^{-7}	−1.4018619
1.5533443	−1.0856646	0.0668194	8.194×10^{-8}	−1.0856644
1.8285469	−1.0503164	0.0747445	4.847×10^{-7}	−1.0503163
2.9622853	−1.0053315	0.0125716	1.018×10^{-10}	−1.0053317
3.0000000	−1.0049451	0.0125716	1.018×10^{-10}	−1.0049452

d)

t	w	h	σ	$y(t)$
0.0570973	0.3328508	0.0570973	3.783×10^{-7}	0.3328508
0.9553711	0.2248909	0.0989115	5.382×10^{-7}	0.2248912
1.4133784	0.1397198	0.0623612	5.025×10^{-7}	0.1397201
1.9342009	0.0616207	0.0219330	4.718×10^{-9}	0.0616209
2.0000000	0.0543453	0.0219330	4.718×10^{-9}	0.0543455

6. The following tables list representative results from the Adams Variable Step-Size Predictor-Corrector method, including the values at each step size change.

a)

t	w	h	$y(t)$
1.8067059	−0.0523479	0.0342520	−0.0523479
2.0408769	−0.3330220	0.0286589	−0.3330207
2.3529653	−0.3587622	0.0254998	−0.3587605
2.9346145	−0.1414994	0.0206531	−0.1415007
3.1412570	−0.3227702	0.0001119	−0.3227704
3.1415927	−0.3230388	0.0001119	−0.3230390

b)

t	w	h	$y(t)$
1.5924693	0.7328084	0.0216730	0.7328084
1.7225071	1.3503009	0.0216730	1.2414825
1.8525450	2.0736417	0.0216730	1.8172666
1.9609098	2.8023162	0.0216730	2.4939426
2.0692747	3.6984817	0.0216730	3.3159234
2.1559666	4.5833276	0.0216730	4.5833257
	$HMIN = 0.02$ exceeded		

c)

t	w	h	$y(t)$
2.8132319	2.0195992	0.0949501	2.0195992
4.0023940	2.3425675	0.1447110	2.3425696
4.9036029	2.5953904	0.1776536	2.5953920
5.4365637	2.7385881	0.1776536	2.7385894

d)

t	w	h	$y(t)$
0.0659375	0.0681114	0.0659375	0.0681114
0.9061680	1.2951225	0.0489803	1.2951247
1.5467518	2.3792502	0.0528206	2.3792499
2.3263125	3.3174224	0.0928934	3.3174244
2.9824245	3.9821533	0.0058585	3.9821504
3.0000000	3.9997558	0.0058585	3.9997532

7. $i(2) = 8.69329$

EXERCISE SET 5.7 (*Page* 193)

1. The Runge-Kutta for Systems method gives the results in the following tables.

a)

i	t_i	w_{1i}	w_{2i}
1	0.5	5.7864583	8.8203125
2	1.0	48.263658	50.104513
3	1.5	342.55321	343.67019
4	2.0	2400.7289	2401.4066

b)

i	t_i	w_{1i}	w_{2i}
1	0.2	2.1203658	1.5069919
2	0.4	4.4412278	3.2422402
3	0.6	9.7391333	8.1634170
4	0.8	22.676560	21.343528
5	1.0	55.661181	56.030503

c)

i	t_i	w_{1i}	w_{2i}
5	0.5	0.9567139	−1.083820
10	1.0	1.306544	−0.8329536
15	1.5	1.344167	−0.5698033
20	2.0	1.143324	−0.3693632

d)

i	t_i	w_{1i}	w_{2i}	w_{3i}
2	0.2	2.997466	−0.03733361	0.7813973
5	0.5	2.963536	−0.2083733	0.3981578
7	0.7	2.905641	−0.3759597	0.1206261
10	1.0	2.749648	−0.6690454	−0.3011653

e)

i	t_i	w_{1i}	w_{2i}	w_{3i}
2	0.2	1.381651	1.007999	−0.6183311
5	0.5	1.907526	1.124998	−0.09090565
7	0.7	2.255029	1.342997	0.2634397
10	1.0	2.832112	1.999997	0.8821206

f)

i	t	w_{1i}	w_{2i}	w_{3i}
2	0.2	7.1615826	−8.9381975	−0.9074132
5	0.5	29.508292	−23.861700	12.516816
7	0.7	83.531779	−67.363422	54.798414
10	1.0	442.99766	−392.18697	377.31385

2. The Runge-Kutta for Systems method gives the results in the following tables.

a)

i	t_i	w_{1i}	w_{2i}
2	0.200	0.0001535	0.0001535
5	0.500	0.0074297	0.0074303
7	0.700	0.0329962	0.0329980
10	1.000	0.1713222	0.1713288

b)

i	t_i	w_{1i}	w_{2i}
5	0.5	0.71544570	-0.19435011
10	1.0	0.77153962	0.40366236
15	1.5	1.1204202	1.0088608
20	2.0	1.8134261	1.8134371

c)

i	t_i	w_{1i}	w_{2i}
5	1.25	0.9405698	−0.4613824
10	1.50	0.7779721	−0.8190555
15	1.75	0.5425639	−1.038640
20	2.00	0.2725883	−1.091117

d)

t_i	w_{1i}	w_{2i}
1.4	2.4535431	2.4540442
2.0	8.7869931	8.7888431
2.4	15.4977930	15.5011215
3.0	30.3916008	30.3982630

e)

t_i	w_{1i}	w_{2i}
1.25	0.7788474	0.7788475
1.50	1.8171683	1.8171685
1.75	3.2203661	3.2203665
2.00	5.1217027	5.1217034

f)

i	t_i	w_{1i}	w_{2i}	w_{3i}
5	1.0	3.731627	4.181249	4.457219
10	2.0	11.31425	12.50243	13.75296
15	3.0	34.04396	37.36869	40.73623

g)

i	t_i	w_{1i}	w_{2i}	w_{3i}
2	1.1	−1.111110	−1.234565	−2.743482
6	1.3	−1.428566	−2.040798	−5.830883
12	1.6	−2.499944	−6.249534	−31.24937
18	1.9	−9.967667	−98.91652	−1990.752

h)

t_i	w_{1i}	w_{2i}
1.200	0.2727376	0.2727379
1.500	1.0884908	1.0884926
1.700	2.0435321	2.0435364
2.000	4.3615668	4.3615778

3. To approximate the solution of the mth–order system of the first–order intial–value problems

$$u_j' = f_j(t, u_1, u_2, \ldots, u_m), \quad j = 1, 2, \ldots, m \quad \text{for} \quad a \leq t \leq b, u_j(a) = \alpha_j, \quad j = 1, 2, \ldots, m$$

at $(n+1)$ equally spaced numbers in the interval $[a, b]$, construct the following algorithm.

INPUT endpoints a, b; number of equations m; integer N; initial conditions $\alpha_1, \ldots, \alpha_m$.
OUTPUT approximations $w_{i,j}$ to $u_j(t_i)$ at t_i.

Step 1 Set $h = (b-a)/N$;
$t_0 = a$.
Step 2 For $j = 1, 2, \ldots, m$ set $w_{0,j} = \alpha_j$
Step 3 OUTPUT $(t_0, w_{0,1}, w_{0,2}, \ldots, w_{0,m})$.
Step 4 For $i = 1, 2, \ldots, N$ do steps 5–11.
Step 5 For $j = 1, 2, \ldots, m$ set
$k_{1,j} = hf_j(t_{i-1}, w_{i-1,1}, \ldots, w_{i-1,m})$
Step 6 For $j = 1, 2, \ldots, m$ set
$k_{2,j} = hf_j(t + \frac{h}{2}, w_{i-1,1} + \frac{1}{2}k_{1,1}, w_{i-1,2} + \frac{1}{2}k_{1,2}, \ldots, w_{i-1,m} + \frac{1}{2}k_{1,m})$.
Step 7 For $j = 1, 2, \ldots, m$ set
$k_{3,j} = hf_j(t + \frac{h}{2}, w_{i-1,1} + \frac{1}{2}k_{2,1}, w_{i-1,2} + \frac{1}{2}k_{2,2}, \ldots, w_{i-1,m} + \frac{1}{2}k_{2,m})$.
Step 8 For $j = 1, 2, \ldots, m$ set
$k_{4,j} = hf_j(t + h, w_{i-1,1} + k_{3,1}, w_{i-1,2} + k_{3,2}, \ldots, w_{i-1,m} + k_{3,m})$.
Step 9 For $j = 1, 2, \ldots, m$ set
$w_{i,j} = w_{i-1,j} + (k_{1,j} + 2k_{2,j} + 2k_{3,j} + k_{4,j})/6$.
Step 10 Set $t_i = a + ih$.
Step 11 OUTPUT $(t_i, w_{i,1}, w_{i,2}, \ldots, w_{i,m})$.
Step 12 For $i = 4, \ldots, N$ do Steps 13–16.
Step 13 Set $t_i = a + ih$.
Step 14 For $j = 1, 2, \ldots, m$ set

$$w_{i,j} = w_{i-1,j} + h[55f_j(t_{i-1}, w_{i-1,1}, \dots, w_{i-1,m}) - 59f_j(t_{i-2}, w_{i-2,1}, \dots, w_{i-2,m}) + 37f_j(t_{i-3}, w_{i-3,1}, \dots, w_{i-3,m}) - 9f_j(t_{i-4}, w_{i-4,1}, \dots, w_{i-4,m})]/24;$$

Step 15 For $j = 1, 2, \dots, m$ set

$$w_{i,j} = w_{i-1,j} + h[9f_j(t_i, w_{i,1}, \dots, w_{i,m}) + 19f_j(t_{i-1}, w_{i-1,1}, \dots, w_{i-1,m}) - 5f_j(t_{i-2}, w_{i-2,1}, \dots, w_{i-2,m}) + f_j(t_{i-3}, w_{i-3,1}, \dots, w_{i-3,m})]/24;$$

Step 16 OUTPUT $(t_i, w_{i,1}, w_{i,2}, \dots, w_{i,m})$.

Step 17 STOP

4. The Adams fourth-order predictor-corrector method for systems applied to the problems in Exercise 2 gives the results in the following tables.

a)

t_i	w_{1i}	w_{2i}	w_{3i}	w_{4i}
0.0	−1.0000000	−1.0000000	4.0000000	4.0000000
0.5	5.7864583	6.1759948	8.8203125	9.2086481
1.0	48.2636583	53.8623912	50.1045125	55.7017884
1.5	342.5532093	402.9825332	343.6701860	404.0981840
2.0	2137.2310485	2980.6873165	2386.8156059	2981.3639929

b)

t_i	w_{1i}	w_{2i}	w_{3i}	w_{4i}
0.0	1.0000000	1.0000000	1.0000000	1.0000000
0.2	2.1203658	2.1250084	1.5069919	1.5115874
0.4	4.4412278	4.4651196	3.2422402	3.2659853
0.6	9.7391333	9.8323587	8.1634170	8.2562955
0.8	22.5267321	23.0026394	21.4221246	21.6688767
1.0	54.9751325	56.7374827	56.1369150	57.1053621

c)

i	t_i	w_{1i}	w_{2i}
2	0.2	0.49548179	−1.1155433
4	0.4	0.83141276	−1.1122947
6	0.6	1.05990410	−1.0440282
8	0.8	1.12122424	−0.94415889
10	1.0	1.30656880	−0.83291735

d)

i	t_i	w_{1i}	w_{2i}	w_{3i}
2	0.2	2.997466	−0.03733361	0.7813973
5	0.5	2.963537	−0.2083736	0.3981572
7	0.7	2.905642	−0.3759606	0.1206247
10	1.0	2.749649	−0.6690474	−0.3011681

e)

i	t_i	w_{1i}	w_{2i}	w_{3i}
2	0.2	1.381651	1.007999	−0.6183311
5	0.5	1.907525	1.124998	−0.09090662
7	0.7	2.255025	1.342997	0.2634382
10	1.0	2.832104	1.999996	0.8821178

f)

t_i	w_{1i}	w_{2i}	w_{3i}
0.0	3.0000000	3.0000000	
0.1	4.6210619	−7.6818883	−1.7287952
0.2	7.1615826	−8.9381975	−0.9074132
0.3	11.2613798	−11.2826255	0.9731908
0.4	18.0396391	−15.6666904	4.8657093
0.5	29.4884956	−23.8531072	12.5269187
0.6	49.1594277	−39.0848095	27.1983282
0.7	83.4197693	−67.3149063	54.8352532
0.8	143.7310704	−119.4492701	106.3524955
0.9	250.7849009	−215.4365214	201.7204580
1.0	442.0182550	−391.7244979	377.4327495

5. The Adams fourth-order predictor-corrector method for systems applied to the problems in Exercise 2 gives the results in the following tables.

a)

i	t_i	w_{1i}	$y(t_i)$
2	0.2	0.0001535	0.0001535
5	0.5	0.0074311	0.0074303
7	0.7	0.0330013	0.0329980
10	1.0	0.1713402	0.1713288

b)

i	t_i	w_{1i}	$y(t_i)$
5	0.5	0.71544570	0.71544564
10	1.0	0.77153962	0.77154031
15	1.5	1.1204202	1.1204222
20	2.0	1.8134261	1.8134302

c)

i	t_i	w_{1i}	$y(t_i)$
5	1.25	0.9405698	0.94056987
10	1.50	0.7779721	0.7779237
15	1.75	0.5425639	0.54256418
20	2.00	0.2725883	0.27258872

d)

i	t_i	w_{1i}	$y(t_i)$
2	1.4	2.4535431	2.4540442
5	2.0	8.7887694	8.7888431
7	2.4	15.5003305	15.5011215
10	3.0	30.3955698	30.3982630

e)

i	t_i	w_i	$y(t_i)$
5	1.25	0.7788478	0.7788475
10	1.50	1.8171693	1.8171685
15	1.75	3.2203676	3.2203665
20	2.00	5.1217050	5.1217034

f)

i	t_i	w_{1i}	$y(t_i)$
5	1.0	3.731627	3.7317044
10	2.0	11.31425	11.314529
15	3.0	34.04396	34.045171

g)

i	t_i	w_{1i}	$y(t_i)$
2	1.1	−1.111110	$-1.\overline{1}$
6	1.3	−1.428566	$-1.\overline{428571}$
12	1.6	−2.499944	−2.5
18	1.9	−9.967667	−10.0

h)

i	t_i	w_{1i}	$y(t_i)$
2	1.2	0.2727376	0.2727379
5	1.5	1.0884792	1.0884926
7	1.7	2.0435228	2.0435364
10	2.0	4.3615685	4.3615778

6. The predicted number of prey, x_{1i}, and predators, x_{2i}, are given in the following table.

i	t_i	x_{1i}	x_{2i}
2	1.0	8716	1435
4	2.0	7907	2120
6	3.0	6666	2813

A stable solution is $x_1 = 833.\overline{3}$ and $x_2 = 1500$.

7. The predicted number of prey, x_{1i}, and predators, x_{2i}, are given in the following table.

i	t_i	w_{1i}	w_{2i}
6	1.2	2211	11429
12	2.4	175	17492
18	3.6	2	19704

A stable solution is $x_1 = 8000$ and $x_2 = 4000$.

EXERCISE SET 5.8 (*Page* 198)

1. Euler's method gives the results in the following tables.

a)

i	t_i	w_i	$y(t_i)$
2	0.2	0.0271828	0.4493290
5	0.5	0.0000272	0.0301974
7	0.7	0.0000003	0.0049916
10	1.0	0.0000000	0.0003355

b)

i	t_i	w_i	$y(t_i)$
2	0.4	1.1200000	0.4815244
5	1.0	0.8444800	1.0006709
7	1.4	1.3440128	1.4000273
10	2.0	2.0120932	2.0000002

c)

i	t_i	w_i	$y(t_i)$
2	0.2	0.3733333	0.0461052
5	0.5	−0.0933333	0.2500151
7	0.7	0.1466667	0.4900003
10	1.0	1.3333333	1.0000000

d)

i	t_i	w_i	$y(t_i)$
2	0.50	14.005208	0.4794709
4	1.00	229.56899	0.8414710
6	1.50	3684.7269	0.9974950
8	2.00	58970.544	0.9092974

e)

i	t_i	w_i	$y(t_i)$
2	1.2	0.9882272	0.9805185
5	1.5	0.8868914	0.8778068
7	1.7	0.7733902	0.7649502
10	2.0	0.5469651	0.5403338

f)

i	t_i	w_i	$y(t_i)$
2	0.2	6.1282588	1.0000000
5	0.5	−378.25743	1.0000000
7	0.7	−6052.0627	1.0000000
10	1.0	387332.00	1.0000000

2. The Runge-Kutta fourth order method gives the results in the following tables.

a)

t_i	w_i	$y(t_i)$	$\lvert w_i - y(t_i)\rvert$
0.2	0.4588119	0.4493290	9.48290×10^{-3}
0.5	0.0318159	0.0301974	1.61856×10^{-3}
0.7	0.0053701	0.0049916	3.78539×10^{-4}
1.0	0.0003724	0.0003355	3.69250×10^{-5}

b)

t_i	w_i	$y(t_i)$	$\|w_i - y(t_i)\|$
0.4	0.5462323	0.4815244	6.47079×10^{-2}
1.0	1.0028911	1.0006709	2.22018×10^{-3}
1.4	1.4002114	1.4000273	1.84038×10^{-4}
2.0	2.0000042	2.0000002	3.95417×10^{-6}

c)

t_i	w_i	$y(t_i)$	$\|w_i - y(t_i)\|$
0.2	0.0792593	0.0461052	3.31540×10^{-2}
0.5	0.2538615	0.2500151	3.84632×10^{-3}
0.7	0.4926513	0.4900003	2.65100×10^{-3}
1.0	1.0025056	1.0000000	2.50560×10^{-3}

d)

t_i	w_i	$y(t_i)$	$\|w_i - y(t_i)\|$
0.50	188.3082146	0.4794709	1.87829×10^{2}
1.00	35296.6783565	0.8414710	3.52958×10^{4}
1.50	6632737.2399876	0.9974950	6.63274×10^{6}
2.00	1246413200.4514600	0.9092974	1.24641×10^{9}

e)

t_i	w_i	$y(t_i)$	$\|w_i - y(t_i)\|$
1.20	0.9804973	0.9805185	2.11925×10^{-5}
1.50	0.8777787	0.8778068	2.80954×10^{-5}
1.70	0.7649232	0.7649502	2.69196×10^{-5}
2.00	0.5403122	0.5403338	2.15420×10^{-5}

f)

t_i	w_i	$y(t_i)$	$\|w_i - y(t_i)\|$
0.20	−215.7458639	1.0000000	2.16746×10^{2}
0.50	−555750.0076938	1.0000000	5.55751×10^{5}
0.70	−104435653.7889370	1.0000000	1.04436×10^{8}
1.00	−269031268010.8340000	1.0000000	2.69031×10^{11}

3. Adams Fourth-Order Predictor-Corrector method gives the results in the following tables.

a)

i	t_i	w_i	$y(t_i)$
2	0.2	0.4588119	0.4493290
5	0.5	−0.0112813	0.0301974
7	0.7	0.0013734	0.0049916
10	1.0	0.0023604	0.0003355

b)

i	t_i	w_i	$y(t_i)$
2	0.4	0.5462323	0.4815244
5	1.0	0.6661010	1.0006709
7	1.4	1.7300815	1.4000273
10	2.0	1.6477554	2.0000002

c)

i	t_i	w_i	$y(t_i)$
2	0.2	0.0792593	0.0461052
5	0.5	0.1554027	0.2500151
7	0.7	0.5507445	0.4900003
10	1.0	0.7278557	1.0000000

d)

i	t_i	w_i	$y(t_i)$
2	0.50	1.8830821×10^2	0.4794709
4	1.00	3.8932032×10^4	0.8414710
6	1.50	9.0736073×10^6	0.9974950
8	2.00	2.1157413×10^8	0.9092974

e)

i	t_i	w_i	$y(t_i)$
2	1.2	0.9804973	0.9805185
5	1.5	0.8778061	0.8778068
7	1.7	0.7649501	0.7649502
10	2.0	0.5403329	0.5403338

f)

i	t_i	w_i	$y(t_i)$
2	0.2	-2.1574586×10^2	1.0000000
5	0.5	-6.8263697×10^5	1.0000000
7	0.7	-1.5917274×10^8	1.0000000
10	1.0	$-5.6675117 \times 10^{11}$	1.0000000

4. The Trapezoidal method gives the results in the following tables.

a)

t_i	w_i	$y(t_i)$	Error	Iterations
0.20	0.3910964	0.4493290	0.0582325	2
0.50	0.0213436	0.0301974	0.0088538	2
0.70	0.0030708	0.0049916	0.0019208	2
1.00	0.0001676	0.0003355	0.0001679	2

b)

t_i	w_i	$y(t_i)$	Error	Iterations
0.40	0.4246914	0.4815244	0.0568330	2
1.00	1.0000339	1.0006709	0.0006371	2
1.40	1.4000004	1.4000273	0.0000269	2
2.00	2.0000000	2.0000002	0.0000002	2

c)

t_i	w_i	$y(t_i)$	Error	Iterations
0.20	0.0400000	0.0461052	0.0061052	2
0.50	0.2500000	0.2500151	0.0000151	2
0.70	0.4900000	0.4900003	0.0000003	2
1.00	1.0000000	1.0000000	0.0000000	2

d)

t_i	w_i	$y(t_i)$	Error	Iterations
0.50	0.6629113	0.4794709	0.1834404	2
1.25	0.9344277	0.9489846	0.0145569	2
1.75	0.9813633	0.9839859	0.0026226	2

e)

t_i	w_i	$y(t_i)$	Error	Iterations
1.20	0.9805606	0.9805185	0.0000421	2
1.50	0.8778685	0.8778068	0.0000617	2
1.70	0.7650334	0.7649502	0.0000833	2
2.00	0.5404520	0.5403338	0.0001182	2

f)

t_i	w_i	$y(t_i)$	Error	Iterations
0.20	−1.0756831	1.0000000	2.0756831	4
0.50	−0.9786836	1.0000000	1.9786836	4
0.70	−0.9904641	1.0000000	1.9904641	3
1.00	−1.0028446	1.0000000	2.0028446	3

5. $p(50) \approx 0.10421$

CHAPTER 6 DIRECT METHODS FOR SOLVING LINEAR SYSTEMS

EXERCISE SET 6.2 (*Page* 208)

1. **a)** Intersecting lines whose solution is $x_1 = x_2 = 1$.

b) Intersecting lines whose solution is $x_1 = x_2 = 0$.

c) One line, so there is an infinite number of solutions with $x_2 = \frac{3}{2} - \frac{1}{2}x$.

d) Parallel lines, so there is no solution.

e) One line, so there is an infinite number of solutions with $x_2 = -\frac{1}{2}x_1$.

f) Intersecting lines whose solution is $x_1 = 5$ and $x_2 = 3$.

g) Three lines in the plane that do not intersect at a common point.

h) Three lines in the plane that do not intersect at a common point.

i) Intersecting lines whose solution is $x_1 = \frac{2}{7}$ and $x_2 = -\frac{11}{7}$.

j) Two planes in space which intersect in a line with $x_1 = -\frac{5}{4}x_2$ and $x_3 = \frac{3}{2}x_2 + 1$.

2. **a)** $x_1 = 1.0, x_2 = -0.98, x_3 = 2.9$ **b)** $x_1 = 1.0, x_2 = -1.0, x_3 = 3.0$

c) $x_1 = 1.1, x_2 = -1.1, x_3 = 2.9$ **d)** $x_1 = 1.0, x_2 = -1.0, x_3 = 3.0$

3. Gaussian elimination gives the following solutions:

a) $x_1 = 1.1875, x_2 = 1.8125, x_3 = 0.875$ with one row interchange required;

b) $x_1 = 0.75, x_2 = 0.5, x_3 = -0.125$ with one row interchange required;

c) $x_1 = -1, x_2 = 0, x_3 = 1$ with no interchange required;

d) $x_1 = 1, x_2 = 2, x_3 = -1$ with no interchange required;

e) $x_1 = 1.5, x_2 = 2, x_3 = -1.2, x_4 = 3$ with no interchange required;

f) $x_1 = \frac{22}{9}, x_2 = -\frac{4}{9}, x_3 = \frac{4}{3}, x_4 = 1$ with one row interchange required;

g) no solution;

h) $x_1 = -1, x_2 = 2, x_3 = 0, x_4 = 1$ with one row interchange required.

4. Gaussian elimination with single precision arithmetic gives the following solutions:

a) $x_1 = -227.0769, x_2 = 476.9231, x_3 = -177.6923$

b) $x_1 = 1.001291, x_2 = 1, x_3 = 1.00155$

c) $x_1 = x_2 = x_3 = x_4 = 1$

d) $x_1 = -0.03174600, x_2 = 0.5952377, x_3 = -2.380951, x_4 = 2.777777$

e) $x_1 = 1.918129, x_2 = 1.964912, x_3 = -0.9883041, x_4 = -3.192982, x_5 = -1.134503$

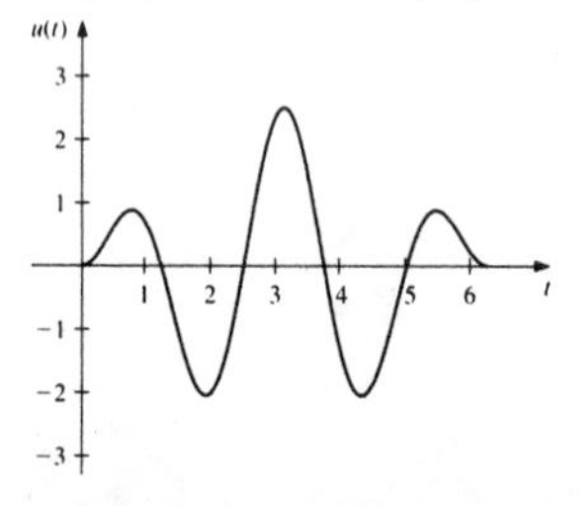

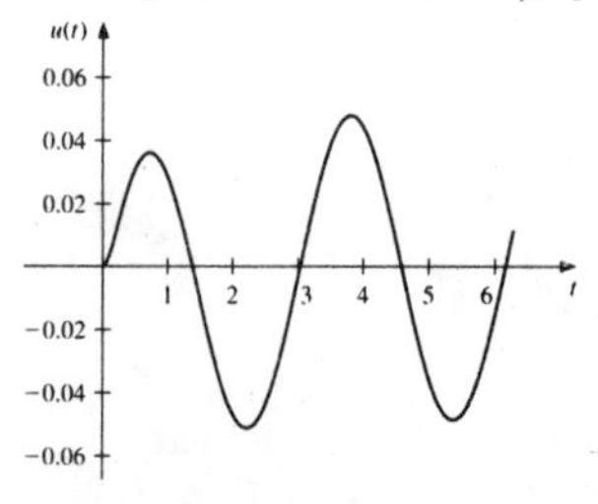

f) $x_1 = 1.956982, x_2 = -2.548580, x_3 = -2.831097, x_4 = 2.421163, x_5 = -0.1054987$

5. a) There is sufficient food to satisfy the average daily consumption.

b) We could add 200 of species 1, or 150 of species 2, or 100 of species 3, or 100 of species 4.

c) Assuming none of the increases indicated in part (b) was selected, species 2 could be increased by 650, or species 3 could be increased by 150, or species 4 could be increased by 150.

d) Assuming none of the increases indicated in parts (b) or (c) were selected, species 3 could be increased by 150, or species 4 could be increased by 150.

6. a) In component form:

$$(a_{11}x_1 - b_{11}y_1 + a_{12}x_2 - b_{12}y_2) + (b_{11}x_1 + a_{11}y_1 + b_{12}x_2 + a_{12}y_2)i = c_1 + id_1$$

$$(a_{21}x_1 - b_{21}y_1 + a_{22}x_2 - b_{22}y_2) + (b_{21}x_1 + a_{21}y_1 + b_{22}x_2 + a_{22}y_2)i = c_2 + id_2$$

which yields

$$\begin{aligned} a_{11}x_1 + a_{12}x_2 - b_{11}y_1 - b_{12}y_2 &= c_1 \\ b_{11}x_1 + b_{12}x_2 + a_{11}y_1 + a_{12}y_2 &= d_1 \\ a_{21}x_1 + a_{22}x_2 - b_{21}y_1 - b_{22}y_2 &= c_2 \\ b_{21}x_1 + b_{22}x_2 + a_{21}y_1 + a_{22}y_2 &= d_2 \end{aligned}$$

b) The system

$$\begin{bmatrix} 1 & 3 & 2 & -2 \\ -2 & 2 & 1 & 3 \\ 2 & 4 & -1 & -3 \\ 1 & 3 & 2 & 4 \end{bmatrix} \begin{bmatrix} x_1 \\ x_2 \\ y_1 \\ y_2 \end{bmatrix} = \begin{bmatrix} 5 \\ 2 \\ 4 \\ -1 \end{bmatrix}$$

has the solution $x_1 = -1.2, x_2 = 1, y_1 = 0.6$, and $y_2 = -1$.

7. a) For the Trapezoidal rule $m = n = 1$, $x_0 = 0$, $x_1 = 1$ so that for $i = 0$ and 1, we have

$$\begin{aligned} u(x_i) &= f(x_i) + \int_0^1 K(x_i, t)u(t)\,dt \\ &= f(x_i) + \frac{1}{2}\Big[K(x_i, 0)u(0) + K(x_i, 1)u(1)\Big]. \end{aligned}$$

Substituting for x_i gives the desired equations.

b) We have $n = 4$, $h = \frac{1}{4}$, $x_0 = 0$, $x_1 = \frac{1}{4}$, $x_2 = \frac{1}{2}$, $x_3 = \frac{3}{4}$, $x_4 = 1$ so

$$\begin{aligned} u(x_i) =& f(x_i) + \frac{h}{2}\left[K(x_i,0)u(0) + K(x_i,1)u(1) + 2K\left(x_i,\frac{1}{4}\right)u\left(\frac{1}{4}\right)\right. \\ &\left. + 2K\left(x_i,\frac{1}{2}\right)u\left(\frac{1}{2}\right) + 2K\left(x_i,\frac{3}{4}\right)u\left(\frac{3}{4}\right)\right] \end{aligned}$$

for $i = 0, 1, 2, 3, 4$. This gives

$$u(x_i) = x_i^2 + \frac{1}{8}[e^{x_i}u(0) + e^{|x_i-1|}u(1) + 2e^{|x_i-\frac{1}{4}|}u\left(\frac{1}{4}\right) + 2e^{|x_i-\frac{1}{2}|}u\left(\frac{1}{2}\right) + 2e^{|x_i-\frac{3}{4}|}u\left(\frac{3}{4}\right)]$$

for each $i = 1, \ldots, 4$. The 5 by 5 linear system has solution $u(0) = -1.154255$, $u(\frac{1}{4}) = -0.9093298$, $u(\frac{1}{2}) = -0.7153145$, $u(\frac{3}{4}) = -0.5472949$, and $u(1) = -0.3931261$.

c) The Composite Simpson's rule gives

$$\begin{aligned} \int_0^1 K(x_i,t)u(t)\,dt =& \frac{h}{3}\left[K(x_i,0)u(0) + 2K\left(x_i,\frac{1}{2}\right)u\left(\frac{1}{2}\right) + 4K\left(x_i,\frac{1}{4}\right)u\left(\frac{1}{4}\right)\right. \\ &\left. 4K\left(x_i,\frac{3}{4}\right)u\left(\frac{3}{4}\right) + K(x_i,1)u(1)\right] \end{aligned}$$

which results in the linear equations

$$u(x_i) = x_i^2 + \frac{1}{12}[e^{x_i}u(0) + e^{|x_i-1|}u(1) + 2e^{|x_i-\frac{1}{2}|}u\left(\frac{1}{2}\right) + 4e^{|x_i-\frac{1}{4}|}u\left(\frac{1}{4}\right) + 4e^{|x_i-\frac{3}{4}|}u\left(\frac{3}{4}\right)].$$

The 5 by 5 linear system has solutions $u(0) = -1.234286$, $u(\frac{1}{4}) = -0.9507292$, $u(\frac{1}{2}) = -0.7659400$, $u(\frac{3}{4}) = -0.5844737$, and $u(1) = -0.4484975$.

EXERCISE SET 6.3 (*Page 215*)

1. Gaussian elimination with three-digit chopping arithmetic gives the following results.

a) $x_1 = 30.0, x_2 = 0.990$ **b)** $x_1 = 1.00, x_2 = 9.98$

c) $x_1 = -40.6, x_2 = -0.125, x_3 = 0.143$ **d)** $x_1 = 9.33, x_2 = 0.492, x_3 = -9.61$

e) $x_1 = -0.102, x_2 = 1.38, x_3 = 2.42$ **f)** $x_1 = 57.8, x_2 = -285, x_3 = 259$

g) $x_1 = 0.198, x_2 = 0.0154, x_3 = -0.0156, x_4 = -0.716$

h) $x_1 = 0.828, x_2 = -3.32, x_3 = 0.153, x_4 = 4.91$

2. Gaussian elimination with three-digit rounding arithmetic gives the following results.

a) $x_1 = -10.0, x_2 = 1.01$ **b)** $x_1 = 1.00, x_2 = 10.0$

c) $x_1 = -40.6, x_2 = -0.125, x_3 = 0.143$ **d)** $x_1 = 8.95, x_2 = 0.497, x_3 = -1.96$

e) $x_1 = -0.113, x_2 = 1.40, x_3 = 2.42$ **f)** $x_1 = 51.1, x_2 = -246, x_3 = 222$

g) $x_1 = 0.176, x_2 = 0.0119, x_3 = -0.0200, x_4 = -1.12$

h) $x_1 = 0.799, x_2 = -3.12, x_3 = 0.151, x_4 = 4.56$

3. Gaussian elimination with maximal column pivoting and three-digit chopping arithmetic gives the following results.

a) $x_1 = 10.0, x_2 = 1.00$ **b)** $x_1 = 1.00, x_2 = 9.98$

c) $x_1 = -0.160, x_2 = 9.98, x_3 = 0.142$ **d)** $x_1 = 9.33, x_2 = 0.492, x_3 = -9.61$

e) $x_1 = -0.102, x_2 = 1.38, x_3 = 2.42$ **f)** $x_1 = 60.1, x_2 = -298, x_3 = 273$

g) $x_1 = 0.172, x_2 = 0.0131, x_3 = -0.208, x_4 = -1.23$

h) $x_1 = 0.777, x_2 = -3.10, x_3 = 0.161, x_4 = 4.50$

4. Gaussian elimination with maximal column pivoting and three-digit rounding arithmetic gives the following results.

a) $x_1 = 10.0, x_2 = 1.00$ **b)** $x_1 = 1.00, x_2 = 10.0$

c) $x_1 = 0.00, x_2 = 10.0, x_3 = 0.143$ **d)** $x_1 = 8.95, x_2 = 0.497, x_3 = -1.96$

e) $x_1 = -0.113, x_2 = 1.40, x_3 = 2.42$ **f)** $x_1 = 51.1, x_2 = -246, x_3 = 222$

g) $x_1 = 0.178, x_2 = 0.0127, x_3 = -0.0204, x_4 = -1.16$

h) $x_1 = 0.845, x_2 = -3.37, x_3 = 0.182, x_4 = 5.07$

5. Gaussian elimination with scaled column pivoting and three-digit chopping arithmetic gives the following results.

a) $x_1 = 10.0, x_2 = 1.00$ **b)** $x_1 = 1.00, x_2 = 9.98$

c) $x_1 = -0.160, x_2 = 9.98, x_3 = 0.142$ **d)** $x_1 = 0.987, x_2 = 0.500, x_3 = -0.997$

e) $x_1 = -0.102, x_2 = 1.38, x_3 = 2.42$ **f)** $x_1 = 60.1, x_2 = -298, x_3 = 273$

g) $x_1 = 0.170, x_2 = 0.0127, x_3 = -0.0217, x_4 = -1.28$

h) $x_1 = 0.837, x_2 = -3.31, x_3 = 0.158, x_4 = 4.92$

6. Gaussian elimination with scaled column pivoting and three-digit rounding arithmetic gives the following results.

a) $x_1 = 10.0, x_2 = 1.00$ **b)** $x_1 = 1.00, x_2 = 10.0$

c) $x_1 = 0.00, x_2 = 10.0, x_3 = 0.143$ **d)** $x_1 = 1.00, x_2 = 0.498, x_3 = -1.00$

e) $x_1 = -0.113, x_2 = 1.40, x_3 = 2.42$ **f)** $x_1 = 52.1, x_2 = -247, x_3 = 222$

g) $x_1 = 0.180, x_2 = 0.0128, x_3 = -0.02000, x_4 = -1.13$

h) $x_1 = 0.783, x_2 = -3.12, x_3 = 0.147, x_4 = 4.53$

7. Gaussian Elimination with Backward Substitution and single precision arithmetic gives the following results.

a) $x_1 = 10.000000, x_2 = 1.0000000$ **b)** $x_1 = 1.0000000, x_2 = 10.000000$

c) $x_1 = 0.0000000, x_2 = 10.000000, x_3 = 0.14285714$

d) $x_1 = 0.99104628, x_2 = 0.49870656, x_3 = -0.99568160$

e) $x_1 = -0.11108022, x_2 = 1.3962863, x_3 = 2.4190803$

f) $x_1 = 54.000044, x_2 = -264.00023, x_3 = 240.00021$

g) $x_1 = 0.17682530, x_2 = 0.012692691, x_3 = -0.020654050, x_4 = -1.1826087$

h) $x_1 = 0.78842555, x_2 = -3.1255199, x_3 = 0.1676848, x_4 = 4.5572988$

8. Gaussian Elimination with Maximal Column Pivoting and single precision arithmetic gives the following results.

a) $x_1 = 10.0, x_2 = 1.00$ **b)** $x_1 = 1.00, x_2 = 10.0$

c) $x_1 = 0.00, x_2 = 10.0, x_3 = 0.143$ **d)** $x_1 = 1.01, x_2 = 0.499, x_3 = -1.00$

e) $x_1 = -0.113, x_2 = 1.40, x_3 = 2.42$ **f)** $x_1 = 51.4, x_2 = -246, x_3 = 221$

g) $x_1 = 0.179, x_2 = 0.0127, x_3 = -0.0203, x_4 = -1.15$

h) $x_1 = 0.874, x_2 = -3.49, x_3 = 0.192, x_4 = 5.33$

9. Gaussian Elimination with Scaled Column Pivoting and single precision arithmetic gives the following results.

a) $x_1 = 10.000000, x_2 = 1.0000000$ **b)** $x_1 = 1.0000000, x_2 = 10.000000$

c) $x_1 = 0.0000000, x_2 = 10.000000, x_3 = 0.14285714$

d) $x_1 = 0.99104628, x_2 = 0.49870656, x_3 = -0.99568160$

e) $x_1 = -0.11108022, x_2 = 1.3962863, x_3 = 2.4190803$

f) $x_1 = 54.000044, x_2 = -264.00023, x_3 = 240.00021$

g) $x_1 = 0.17682530, x_2 = 0.012692691, x_3 = -0.020654050, x_4 = -1.1826087$

h) $x_1 = 0.78842555, x_2 = -3.1255199, x_3 = 0.1676848, x_4 = 4.5572988$

EXERCISE SET 6.4 (*Page* 222)

1. **a)** The matrix is singular.

b) $\begin{bmatrix} -\frac{1}{4} & \frac{1}{4} & \frac{1}{4} \\ \frac{5}{8} & -\frac{1}{8} & -\frac{1}{8} \\ \frac{1}{8} & -\frac{5}{8} & \frac{3}{8} \end{bmatrix}$ **c)** $\begin{bmatrix} \frac{1}{2} & 0 & 0 \\ 0 & -\frac{1}{3} & 0 \\ 0 & 0 & 1 \end{bmatrix}$

d) The matrix is singular. **e)** The matrix is singular. **f)** The matrix is singular.

g) $\begin{bmatrix} \frac{1}{4} & 0 & 0 & 0 \\ -\frac{3}{14} & \frac{1}{7} & 0 & 0 \\ \frac{3}{28} & -\frac{11}{7} & 1 & 0 \\ -\frac{1}{2} & 1 & -1 & 1 \end{bmatrix}$ **h)** $\begin{bmatrix} 1 & 0 & 1 & -1 \\ -1 & \frac{5}{3} & \frac{5}{3} & -1 \\ -1 & \frac{2}{3} & \frac{2}{3} & 0 \\ 0 & -\frac{1}{3} & -\frac{4}{3} & 1 \end{bmatrix}$

i) $\begin{bmatrix} 1 & 0 & 0 & 0 \\ 2 & 1 & 0 & 0 \\ 3 & 4 & 1 & 0 \\ -1 & -3 & 0 & 1 \end{bmatrix}$ **j)** $\begin{bmatrix} 1 & 1 & 2 & 4 \\ 0 & 1 & 1 & 2 \\ 0 & 0 & 1 & 1 \\ 0 & 0 & 0 & 1 \end{bmatrix}$

2. **a)** $\begin{bmatrix} 1 & 0 & 0 \\ 1 & 2 & 0 \\ 9 & 5 & 1 \end{bmatrix}$ **b)** $\begin{bmatrix} 1 & -1 & 2 \\ 2 & -1 & 7 \\ -2 & 1 & -5 \end{bmatrix}$

c) $\begin{bmatrix} 1 & 0 & 0 \\ 2 & 1 & 0 \\ -7 & -2 & 1 \end{bmatrix}$ **d)** $\begin{bmatrix} 6 & -7 & 15 \\ 0 & -1 & 3 \\ 0 & 0 & 6 \end{bmatrix}$

3. The solutions to the linear systems obtained in parts (a) and (b) are, from left to right and top to bottom:
$-\frac{2}{7}, -\frac{13}{14}, -\frac{3}{14}$; $\frac{17}{7}, -\frac{19}{14}, -\frac{41}{14}$; $1, 1, 1$ and $-\frac{1}{7}, \frac{2}{7}, \frac{1}{7}$.

4. Parts a), b), and c): The first system has $x_1 = 3, x_2 = -6, x_3 = -2, x_4 = -1$. The second system $x_1 = 1, x_2 = 1, x_3 = 1, x_4 = 1$.

d) Part (c) requires more work.

5. The determinants of the matricies are:

a) -8 **b)** 14 **c)** 0 **d)** 3

6. We have $\det A = -5.5$, $\det B = -6$, and $\det AB = \det BA = 33$.

7. The result follows from $\det AB = \det A \; \det B$ and the fact that a matrix is nonsingular if and only if its determinant is nonzero.

8. a) Not true. Let

$$A = \begin{bmatrix} 2 & 1 \\ 1 & 0 \end{bmatrix} \quad \text{and} \quad B = \begin{bmatrix} 1 & -1 \\ -1 & 2 \end{bmatrix}. \quad \text{Then} \quad AB = \begin{bmatrix} 1 & 0 \\ 1 & -1 \end{bmatrix}$$

is not symmetric.

b) True. Let A be a nonsingular symmetric matrix. By Theorem 6.12 (d), $(A^{-1})^t = (A^t)^{-1}$. Thus, $(A^{-1})^t = (A^t)^{-1} = A^{-1}$ and A^{-1} is symmetric.

c) Not true. Use the matrices A and B from part (a).

9. a) If $C = AB$, where A and B are lower triangular, then $a_{ik} = 0$ if $k > i$ and $b_{kj} = 0$ if $k < j$. Thus,

$$c_{ij} = \sum_{k=1}^{n} a_{ik} b_{kj} = \sum_{k=j}^{i} a_{ik} b_{kj},$$

which will have the sum zero unless $j \leq i$. Hence C is lower triangular.

b) We have $a_{ik} = 0$ if $k < i$ and $b_{kj} = 0$ if $k > j$. The steps are similar to those in part (a).

c) Let L be a nonsingular lower triangular matrix. To obtain the ith column of L^{-1}, solve n linear systems of the form

$$\begin{bmatrix} l_{11} & 0 & \cdots & \cdots & \cdots & \cdots & \cdots & 0 \\ l_{21} & l_{22} & \ddots & & & & & \vdots \\ \vdots & \vdots & \ddots & \ddots & & & & \vdots \\ \vdots & \vdots & & \ddots & \ddots & & & \vdots \\ l_{i1} & l_{i2} & \cdots & \cdots & l_{ii} & \ddots & & \vdots \\ \vdots & \vdots & & & & \ddots & \ddots & \vdots \\ \vdots & \vdots & & & & & \ddots & 0 \\ l_{n1} & l_{n2} & \cdots & \cdots & \cdots & \cdots & & l_{nn} \end{bmatrix} \begin{bmatrix} x_1 \\ x_2 \\ \vdots \\ \vdots \\ x_i \\ \vdots \\ \vdots \\ x_n \end{bmatrix} = \begin{bmatrix} 0 \\ 0 \\ \vdots \\ 0 \\ 1 \\ 0 \\ \vdots \\ 0 \end{bmatrix},$$

where the 1 appears in the ith position.

10. a)

$$A^2 = \begin{bmatrix} 0 & 2 & 0 \\ 0 & 0 & 3 \\ \frac{1}{6} & 0 & 0 \end{bmatrix}, \quad A^3 = \begin{bmatrix} 1 & 0 & 0 \\ 0 & 1 & 0 \\ 0 & 0 & 1 \end{bmatrix}, \quad A^4 = A, \quad A^5 = A^2, \quad A^6 = I, \ldots$$

b)

	Year 1	Year 2	Year 3	Year 4
Age 1	6000	36000	12000	6000
Age 2	6000	3000	18000	6000
Age 3	6000	2000	1000	6000

c)

$$A^{-1} = \begin{bmatrix} 0 & 2 & 0 \\ 0 & 0 & 3 \\ \frac{1}{6} & 0 & 0 \end{bmatrix}.$$

The i, j-entry is the number of beetles of age i necessary to produce one beetle of age j.

11. a) We have

$$\begin{bmatrix} 7 & 4 & 4 & 0 \\ -6 & -3 & -6 & 0 \\ 0 & 0 & 3 & 0 \\ 0 & 0 & 0 & 1 \end{bmatrix} \begin{bmatrix} 2(x_0 - x_1) + \alpha_0 + \alpha_1 \\ 3(x_1 - x_0) - \alpha_1 - 2\alpha_0 \\ \alpha_0 \\ x_0 \end{bmatrix} = \begin{bmatrix} 2(x_0 - x_1) + 3\alpha_0 + 3\alpha_1 \\ 3(x_1 - x_0) - 3\alpha_1 - 6\alpha_0 \\ 3\alpha_0 \\ x_0 \end{bmatrix}$$

b)

$$B = A^{-1} = \begin{bmatrix} -1 & -\frac{4}{3} & -\frac{4}{3} & 0 \\ 2 & \frac{7}{3} & 2 & 0 \\ 0 & 0 & \frac{1}{3} & 0 \\ 0 & 0 & 0 & 1 \end{bmatrix}$$

EXERCISE SET 6.5 (*Page* 229)

1. a)

$$L = \begin{bmatrix} 1 & 0 & 0 \\ 1.5 & 1 & 0 \\ 1.5 & 1 & 1 \end{bmatrix} \quad \text{and} \quad U = \begin{bmatrix} 2 & -1 & 1 \\ 0 & 4.5 & 7.5 \\ 0 & 0 & -4 \end{bmatrix}$$

b)

$$L = \begin{bmatrix} 1 & 0 & 0 \\ -0.5 & 1 & 0 \\ 2 & 2 & 1 \end{bmatrix} \quad \text{and} \quad U = \begin{bmatrix} 2 & -1.5 & 3 \\ 0 & -0.75 & 3.5 \\ 0 & 0 & -8 \end{bmatrix}$$

c)

$$L = \begin{bmatrix} 1 & 0 & 0 \\ -2.106719 & 1 & 0 \\ 3.067193 & 1.19776 & 1 \end{bmatrix}$$

and

$$U = \begin{bmatrix} 1.012 & -2.132 & 3.104 \\ 0 & -0.3955249 & -0.4737443 \\ 0 & 0 & -8.939133 \end{bmatrix}$$

d) $L = I$ and U is the original matrix.

e)

$$L = \begin{bmatrix} 1 & 0 & 0 & 0 \\ 0.5 & 1 & 0 & 0 \\ 0 & -2 & 1 & 0 \\ 1 & -1.33333 & 2 & 1 \end{bmatrix} \quad \text{and} \quad U = \begin{bmatrix} 2 & 0 & 0 & 0 \\ 0 & 1.5 & 0 & 0 \\ 0 & 0 & 0.5 & 0 \\ 0 & 0 & 0 & 1 \end{bmatrix}$$

f)

$$L = \begin{bmatrix} 1 & 0 & 0 & 0 \\ -1.849190 & 1 & 0 & 0 \\ -0.4596433 & -0.2501219 & 1 & 0 \\ 2.768661 & -0.3079435 & -5.35229 & 1 \end{bmatrix}$$

and

$$U = \begin{bmatrix} 2.175600 & 4.023099 & -2.173199 & 5.196700 \\ 0 & 13.43947 & -4.018660 & 10.80698 \\ 0 & 0 & -0.8929510 & 5.091692 \\ 0 & 0 & 0 & 12.03614 \end{bmatrix}$$

2. a) $x_1 = 1,\ x_2 = 2,\ x_3 = -1$

b) $x_1 = 1.5,\ x_2 = 2,\ x_3 = -1.199998,\ x_4 = 3$

c) $x_1 = 1,\ x_2 = 1,\ x_3 = 1$

d) $x_1 = -3.44744,\ x_2 = 5.57458,\ x_3 = 3.21845$

e) $x_1 = -0.5,\ x_2 = 1.16666,\ x_3 = 1.25$

f) $x_1 = 2.939851,\ x_2 = 0.07067770,\ x_3 = 5.677735,\ x_4 = 4.379812$

3. a)

$$P^tLU = \begin{bmatrix} 0 & 1 & 0 \\ 1 & 0 & 0 \\ 0 & 0 & 1 \end{bmatrix} \begin{bmatrix} 1 & 0 & 0 \\ 0 & 1 & 0 \\ 0 & -\frac{1}{2} & 1 \end{bmatrix} \begin{bmatrix} 1 & 1 & -1 \\ 0 & 2 & 3 \\ 0 & 0 & \frac{5}{2} \end{bmatrix}$$

b)

$$P^tLU = \begin{bmatrix} 1 & 0 & 0 \\ 0 & 0 & 1 \\ 0 & 1 & 0 \end{bmatrix} \begin{bmatrix} 1 & 0 & 0 \\ 2 & 1 & 0 \\ 1 & 0 & 1 \end{bmatrix} \begin{bmatrix} 1 & 2 & -1 \\ 0 & -5 & 6 \\ 0 & 0 & 4 \end{bmatrix}$$

c)

$$P^tLU = \begin{bmatrix} 1 & 0 & 0 & 0 \\ 0 & 1 & 0 & 0 \\ 0 & 0 & 0 & 1 \\ 0 & 0 & 1 & 0 \end{bmatrix} \begin{bmatrix} 1 & 0 & 0 & 0 \\ 2 & 1 & 0 & 0 \\ 3 & 4 & 1 & 0 \\ -1 & -3 & 0 & 1 \end{bmatrix} \begin{bmatrix} 1 & 1 & 0 & 3 \\ 0 & -1 & -1 & -5 \\ 0 & 0 & 3 & 13 \\ 0 & 0 & 0 & -13 \end{bmatrix}$$

d)

$$P^tLU = \begin{bmatrix} 1 & 0 & 0 & 0 \\ 0 & 0 & 0 & 1 \\ 0 & 0 & 1 & 0 \\ 0 & 1 & 0 & 0 \end{bmatrix} \begin{bmatrix} 1 & 0 & 0 & 0 \\ 2 & 1 & 0 & 0 \\ 1 & 0 & 1 & 0 \\ 1 & 0 & 0 & 1 \end{bmatrix} \begin{bmatrix} 1 & -2 & 3 & 0 \\ 0 & 5 & -3 & -1 \\ 0 & 0 & -1 & -2 \\ 0 & 0 & 0 & 1 \end{bmatrix}$$

EXERCISE SET 6.6 (*Page* 236)

1. (*i*) The symmetric matricies are in (a), (b), and (f).
(*ii*) The singular matricies are in (e) and (h).
(*iii*) The strictly diagonally dominant matricies are in (a), (b), (c), and (d).
(*iv*) The positive definite matricies are in (a) and (f).

2. a)

$$L = \begin{bmatrix} 1 & 0 & 0 \\ -\frac{1}{2} & 1 & 0 \\ 0 & -\frac{2}{3} & 1 \end{bmatrix}, D = \begin{bmatrix} 2 & 0 & 0 \\ 0 & \frac{3}{2} & 0 \\ 0 & 0 & \frac{4}{3} \end{bmatrix}$$

b)

$$L = \begin{bmatrix} 1.0 & 0.0 & 0.0 & 0.0 \\ 0.25 & 1.0 & 0.0 & 0.0 \\ 0.25 & -0.45454545 & 1.0 & 0.0 \\ 0.25 & 0.27272727 & 0.076923077 & 1.0 \end{bmatrix}$$

$$D = \begin{bmatrix} 4.0 & 0.0 & 0.0 & 0.0 \\ 0.0 & 2.75 & 0.0 & 0.0 \\ 0.0 & 0.0 & 1.1818182 & 0.0 \\ 0.0 & 0.0 & 0.0 & 1.5384615 \end{bmatrix}$$

c)

$$L = \begin{bmatrix} 1.0 & 0.0 & 0.0 & 0.0 \\ 0.25 & 1.0 & 0.0 & 0.0 \\ -0.25 & -0.27272727 & 1.0 & 0.0 \\ 0.0 & 0.0 & 0.44 & 1.0 \end{bmatrix}$$

$$D = \begin{bmatrix} 4.0 & 0.0 & 0.0 & 0.0 \\ 0.0 & 2.75 & 0.0 & 0.0 \\ 0.0 & 0.0 & 4.5454545 & 0.0 \\ 0.0 & 0.0 & 0.0 & 3.12 \end{bmatrix}$$

d)

$$L = \begin{bmatrix} 1.0 & 0.0 & 0.0 & 0.0 \\ 0.33333333 & 1.0 & 0.0 & 0.0 \\ 0.16666667 & 0.2 & 1.0 & 0.0 \\ -0.16666667 & 0.1 & -0.24324324 & 1.0 \end{bmatrix}$$

$$D = \begin{bmatrix} 6.0 & 0.0 & 0.0 & 0.0 \\ 0.0 & 3.3333333 & 0.0 & 0.0 \\ 0.0 & 0.0 & 3.7 & 0.0 \\ 0.0 & 0.0 & 0.0 & 2.5810811 \end{bmatrix}$$

3. Choleski's method gives the results in the following tables.

a)

$$L = \begin{bmatrix} 1.41423 & 0 & 0 \\ -0.7071069 & 1.224743 & 0 \\ 0 & -0.8164972 & 1.154699 \end{bmatrix}$$

b)

$$L = \begin{bmatrix} 2 & 0 & 0 & 0 \\ 0.5 & 1.658311 & 0 & 0 \\ 0.5 & -0.7537785 & 1.087113 & 0 \\ 0.5 & 0.4522671 & 0.08362442 & 1.240346 \end{bmatrix}$$

c)

$$L = \begin{bmatrix} 2 & 0 & 0 & 0 \\ 0.5 & 1.658311 & 0 & 0 \\ -0.5 & -0.4522671 & 2.132006 & 0 \\ 0 & 0 & 0.9380833 & 1.766351 \end{bmatrix}$$

d)

$$L = \begin{bmatrix} 2.449489 & 0 & 0 & 0 \\ 0.8164966 & 1.825741 & 0 & 0 \\ 0.4082483 & 0.3651483 & 1.923538 & 0 \\ -0.4082483 & 0.1825741 & -0.4678876 & 1.606574 \end{bmatrix}$$

4. The modified factorization method gives the following results.

a) $x_1 = 1, x_2 = -1, x_3 = 0$

b) $x_1 = 0.2, x_2 = -0.2, x_3 = -0.2, x_4 = 0.25$

c) $x_1 = 1, x_2 = 2, x_3 = -1, x_4 = 2$

d) $x_1 = -0.8586387, x_2 = 2.418848, x_3 = -0.9581152, x_4 = -1.272251$

5. The modified Choleski's method gives the following results.

a) $x_1 = 1, x_2 = -1, x_3 = 0$

b) $x_1 = 0.2, x_2 = -0.2, x_3 = -0.2, x_4 = 0.25$

c) $x_1 = 1, x_2 = 2, x_3 = -1, x_4 = -2$

d) $x_1 = -0.85863874, x_2 = 2.4188482, x_3 = -0.95811518, x_4 = -1.2722513$

6. The Crout Factorization method gives the following results.

a) $x_1 = 0.5, x_2 = 0.5, x_3 = 1$

b) $x_1 = -0.9999995, x_2 = 1.999999, x_3 = 1$

c) $x_1 = 1, x_2 = -1, x_3 = 0$

d) $x_1 = -0.09357762, x_2 = 1.587155, x_3 = -1.167431, x_4 = 0.5412842$

7. We have $x_i = 1$ for each $i = 1, \ldots, 10$.

8. a)

$$L = \begin{bmatrix} 1 & 0 & 0 \\ -1 & 1 & 0 \\ 2 & 1 & 1 \end{bmatrix}, \qquad D = \begin{bmatrix} 3 & 0 & 0 \\ 0 & -1 & 0 \\ 0 & 0 & 2 \end{bmatrix}$$

b)

$$L = \begin{bmatrix} 1 & 0 & 0 \\ -2 & 1 & 0 \\ 3 & -1 & 1 \end{bmatrix}, \qquad D = \begin{bmatrix} 3 & 0 & 0 \\ 0 & 1 & 0 \\ 0 & 0 & 0 \end{bmatrix}$$

c)

$$L = \begin{bmatrix} 1 & 0 & 0 & 0 \\ -2 & 1 & 0 & 0 \\ 0 & 2 & 1 & 0 \\ -1 & 1 & 4 & 1 \end{bmatrix}, \qquad D = \begin{bmatrix} -1 & 0 & 0 & 0 \\ 0 & 1 & 0 & 0 \\ 0 & 0 & 1 & 0 \\ 0 & 0 & 0 & -4 \end{bmatrix}$$

d)

$$L = \begin{bmatrix} 1 & 0 & 0 & 0 \\ -1 & 1 & 0 & 0 \\ 2 & 0 & 1 & 0 \\ -2 & 1 & -1 & 1 \end{bmatrix}, \qquad D = \begin{bmatrix} 2 & 0 & 0 & 0 \\ 0 & 1 & 0 & 0 \\ 0 & 0 & 2 & 0 \\ 0 & 0 & 0 & 3 \end{bmatrix}$$

9. Only the matrix in (d) is positive definite.

10. a) Yes.

b) Not necessarily. Consider $\begin{bmatrix} 2 & -1 \\ 3 & 4 \end{bmatrix}$.

c) Not necessarily. Consider $\begin{bmatrix} 2 & 1 \\ 1 & 2 \end{bmatrix}$ and $\begin{bmatrix} -2 & 1 \\ 1 & -2 \end{bmatrix}$.

d) Not necessarily. Consider $\begin{bmatrix} 2 & -1 \\ 3 & 4 \end{bmatrix}$.

e) Not necessarily. Consider $\begin{bmatrix} 2 & 1 \\ 1 & 2 \end{bmatrix}$ and $\begin{bmatrix} 2 & -1 \\ -1 & 2 \end{bmatrix}$.

11. **a)** No, consider $\begin{bmatrix} -1 & 0 \\ 0 & -1 \end{bmatrix}$.

b) Yes, since $A = A^t$.

c) Yes, since $\mathbf{x}^t(A + B)\mathbf{x} = \mathbf{x}^t A\mathbf{x} + \mathbf{x}^t B\mathbf{x}$.

d) Yes, since $\mathbf{x}^t A^2 \mathbf{x} = \mathbf{x}^t A^t A\mathbf{x} = (A\mathbf{x})^t(A\mathbf{x}) \geq 0$ and because A is nonsingular, equality holds only if $\mathbf{x} = \mathbf{0}$.

e) No, consider $A = \begin{bmatrix} 1 & 0 \\ 0 & 1 \end{bmatrix}$ and $B = \begin{bmatrix} 10 & 0 \\ 0 & 10 \end{bmatrix}$.

12. Yes

13. $A = \begin{bmatrix} 1.0 & 0.2 \\ 0.1 & 1.0 \end{bmatrix}$

14. **a)** Mating male i with female j produces offspring with the same wing characteristics as mating male j with female i.

b) No. Consider, for example, $\mathbf{x} = (1, 0, -1)^t$.

Chapter 7 Iterative Methods for Solving Linear Systems

EXERCISE SET 7.2 (*Page* 249)

1. **a)** $||\mathbf{x}||_\infty = 4$ and $||\mathbf{x}||_2 = 5.220153$. **b)** $||\mathbf{x}||_\infty = 4$ and $||\mathbf{x}||_2 = 5.477226$.

c) $||\mathbf{x}||_\infty = 2^k$ and $||\mathbf{x}||_2 = (1+4^k)^{\frac{1}{2}}$.

d) $||\mathbf{x}||_\infty = 4/(k+1)$ and $||\mathbf{x}||_2 = (16/(k+1)^2 + 4/k^4 + k^4e^{-2k})^{\frac{1}{2}}$.

2. **a)** Since $||\mathbf{x}||_1 = \sum_{i=1}^n |x_i| \geq 0$ with equality only if $x_i = 0$ for all i, properties (i) and (ii) hold.

Also,

$$||\alpha\mathbf{x}||_1 = \sum_{i=1}^n |\alpha x_i| = \sum_{i=1}^n |\alpha||x_i| = |\alpha|\sum_{i=1}^n |x_i| = |\alpha|||\mathbf{x}||_1$$

so property (iii) holds.

Finally,

$$||\mathbf{x}+\mathbf{y}||_1 = \sum_{i=1}^n |x_i + y_i| \leq \sum_{i=1}^n (|x_i| + |y_i|) = \sum_{i=1}^n |x_i| + \sum_{i=1}^n |y_i| = ||\mathbf{x}||_1 + ||\mathbf{y}||_1,$$

so property (iv) also holds.

b) (1a) 8.5 (1b) 10 (1c) $|\sin k| + |\cos k| + e^k$ (1d) $4/(k+1) + 2/k^2 + k^2e^{-k}$

3. **a)** We have $\lim_{k\to\infty} \mathbf{x}^{(k)} = (0,0,0)^t$. **b)** We have $\lim_{k\to\infty} \mathbf{x}^{(k)} = (0,1,3)^t$.

c) We have $\lim_{k\to\infty} \mathbf{x}^{(k)} = (0,0,\frac{1}{2})^t$. **d)** We have $\lim_{k\to\infty} \mathbf{x}^{(k)} = (1,-1,1)^t$.

4. **a)** 25 **b)** 16 **c)** 4 **d)** 12

5. **a)** We have $||\mathbf{x}-\hat{\mathbf{x}}||_\infty = 6.67 \times 10^{-4}$ and $||A\hat{\mathbf{x}} - \mathbf{b}||_\infty = 2.06 \times 10^{-4}$.

b) We have $||\mathbf{x}-\hat{\mathbf{x}}||_\infty = 0.33$ and $||A\hat{\mathbf{x}} - \mathbf{b}||_\infty = 0.27$.

c) We have $||\mathbf{x}-\hat{\mathbf{x}}||_\infty = 0.5$ and $||A\hat{\mathbf{x}} - \mathbf{b}||_\infty = 0.3$.

d) We have $||\mathbf{x}-\hat{\mathbf{x}}||_\infty = 6.55 \times 10^{-2}$, and $||A\hat{\mathbf{x}} - \mathbf{b}||_\infty = 0.32$.

6. **a)** 16 **b)** 25 **c)** 4 **d)** 12

EXERCISE SET 7.3 (*Page* 254)

1. **a)** The eigenvalue $\lambda_1 = 3$ has the eigenvector $\mathbf{x}_1 = (1,-1)^t$ and the eigenvalue $\lambda_2 = 1$ has the eigenvector $\mathbf{x}_2 = (1,1)^t$.

b) The eigenvalue $\lambda_1 = \lambda_2 = 1$ has the eigenvector $\mathbf{x} = (1,0)^t$.

c) The eigenvalue $\lambda_1 = \frac{1}{2}$ has the eigenvector $\mathbf{x}_1 = (1,1)^t$ and the eigenvalue $\lambda_2 = -\frac{1}{2}$ has the eigenvector $\mathbf{x}_2 = (1,-1)^t$.

d) The eigenvalue $\lambda_1 = 0$ has the eigenvector $\mathbf{x}_1 = (1,-1)^t$ and the eigenvalue $\lambda_2 = -1$ has the eigenvector $\mathbf{x}_2 = (1,-2)^t$.

e) The eigenvalue $\lambda_1 = \lambda_2 = 3$ has the eigenvectors $\mathbf{x}_1 = (0,0,1)^t$ and $\mathbf{x}_2 = (1,1,0)^t$, and the eigenvalue $\lambda_3 = 1$ has the eigenvector $\mathbf{x}_3 = (1,-1,0)^t$.

f) The eigenvalue $\lambda_1 = 7$ has the eigenvector $\mathbf{x}_1 = (1,4,4)^t$, the eigenvalue $\lambda_2 = 3$ has the eigenvector $\mathbf{x}_2 = (1,2,0)^t$, and the eigenvalue $\lambda_3 = -1$ has the eigenvector $\mathbf{x}_3 = (1,0,0)^t$.

g) The eigenvalue $\lambda_1 = \lambda_2 = 1$ has the eigenvectors $\mathbf{x}_1 = (-1,1,0)^t$ and $\mathbf{x}_2 = (-1,0,1)^t$, and the eigenvalue $\lambda_3 = 5$ has the eigenvector $\mathbf{x}_3 = (1,2,1)^t$.

h) The eigenvalue $\lambda_1 = 3$ has the eigenvector $\mathbf{x}_1 = (-0.408248, 0.408248, 0.816497)^t$, the eigenvalue $\lambda_2 = 4$ has the eigenvector $\mathbf{x}_2 = (0, 0.447214, 0.447214)^t$ and the eigenvalue $\lambda_3 = -2$ has the eigenvector $\mathbf{x}_3 = (0.348743, -0.929981, -0.116248)^t$.

2. **a)** 3 **b)** 1 **c)** 0.5 **d)** 1
e) 3 **f)** 7 **g)** 5 **h)** 4

3. Since

$$A_1^k = \begin{bmatrix} 1 & 0 \\ \frac{2^k-1}{2^{k+1}} & 2^{-k} \end{bmatrix}, \text{ we have } \lim_{k\to\infty} A_1^k = \begin{bmatrix} 1 & 0 \\ \frac{1}{2} & 0 \end{bmatrix}.$$

Also

$$A_2^k = \begin{bmatrix} 2^{-k} & 0 \\ \frac{16k}{2^{k-1}} & 2^{-k} \end{bmatrix}, \text{ so } \lim_{k\to\infty} A_2^k = \begin{bmatrix} 0 & 0 \\ 0 & 0 \end{bmatrix}.$$

4. Only the matrix in (c) is convergent.

5. **a)** 3 **b)** 1 **c)** $-\frac{1}{4}$ **d)** 0
e) 9 **f)** −21 **g)** 5 **h)** −24

6. **a)** 3 **b)** 1.618034 **c)** 0.5 **d)** 3.162278
e) 3 **f)** 8.224257 **g)** 5.203527 **h)** 5.601152

7. **a)** We have the real eigenvalue $\lambda = 1$ with the eigenvector $\mathbf{x} = (6,3,1)^t$.

b) Choose any multiple of the vector $(6,3,1)^t$.

EXERCISE SET 7.4 (*Page* 259)

1. Two iterations of Jacobi's method give the following results.

a) $(0.1428571, -0.3571429, 0.4285714)^t$ **b)** $(0.97, 0.91, 0.74)^t$

c) $(1.4, 2.85, 1.2, 1.5)^t$ **d)** $(0.075, 2.9625, -1.1875, -3.975)^t$

e) $(2.975, -2.65, 1.75, 0.325)^t$ **f)** $(1.325, -1.6, 1.6, 1.675, 2.425)^t$

g) $(-0.5208333, -0.04166667, -0.2166667, 0.4166667)^t$

h) $(0.6875, 1.125, 0.6875, 1.375, 0.5625, 1.375)^t$

2. Two iterations of the Gauss-Seidel method give the following results.

a) $\mathbf{x}^{(2)} = (0.1111111, -0.2222222, 0.6190476)^t$

b) $\mathbf{x}^{(2)} = (0.979, 0.9495, 0.7899)^t$

c) $\mathbf{x}^{(2)} = (1.4, 2.85, 1.2, 1.5)^t$

d) $\mathbf{x}^{(2)} = (0.172, 3.5452, -1.36985, -4.60109)^t$

e) $\mathbf{x}^{(2)} = (2.9375, -1.15625, 0.6125, 0.5734375)^t$

f) $\mathbf{x}^{(2)} = (1.189063, -1.521354, 1.862396, 1.882526, 2.255645)^t$

g) $\mathbf{x}^{(2)} = (-0.625, 0, -0.225, 0.6166667)^t$

h) $\mathbf{x}^{(2)} = (0.6875, 1.546875, 0.7929688, 1.71875, 0.7226563, 1.878906)^t$

3. Jacobi's method gives the following results.

a) $\mathbf{x}^{(10)} = (0.03507839, -0.2369262, 0.6578015)^t$

b) $\mathbf{x}^{(6)} = (0.9957250, 0.9577750, 0.7914500)^t$

c) $\mathbf{x}^{(5)} = (1.95, 2.9, 1.2, 1.5)^t$

d) $\mathbf{x}^{(13)} = (-0.08268421, 3.789723, -1.519298, -4.777234)^t$

e) Does not converge in 25 itereations.

f) $\mathbf{x}^{(12)} = (0.7870883, -1.003036, 1.866048, 1.912449, 1.98571)^t$

g) $\mathbf{x}^{(14)} = (-0.7529267, 0.04078538, -0.2806091, 0.6911662)^t$

h) $\mathbf{x}^{(17)} = (0.9996805, 1.999774, 0.9996805, 1.999840, 0.9995482, 1.999840)^t$

4. The Gauss-Seidel method gives the following results.

a) $\mathbf{x}^{(6)} = (0.03535107, -0.2367886, 0.6577590)^t$

b) $\mathbf{x}^{(4)} = (0.9957475, 0.9578738, 0.7915748)^t$

c) $\mathbf{x}^{(5)} = (1.95, 2.9, 1.2, 1.5)^t$

d) $\mathbf{x}^{(7)} = (-0.08274495, 3.789609, -1.519167, -4.777598)^t$

e) Did not converge in 25 iterations.

f) $\mathbf{x}^{(7)} = (0.7866825, -1.002719, 1.866283, 1.912562, 1.989790)^t$

g) $\mathbf{x}^{(8)} = (-0.7531763, 0.04101049, -0.2807047, 0.6916305)^t$

h) $\mathbf{x}^{(10)} = (0.9998334, 1.999858, 0.9999393, 1.999899, 0.9999142, 1.999963)^t$

EXERCISE SET 7.5 (*Page* 263)

1. Two iterations of the SOR method with $\omega = 1.1$ give the following results.

a) $(0.05410079, -0.2115435, 0.6477159)^t$ **b)** $(0.9876790, 0.9784935, 0.7899328)^t$

c) $(1.6060, 2.9645, 1.2540, 1.4850)^t$

d) $(0.08765735, 3.818345, -1.471937, -4.824042)^t$

e) $(3.055932, -1.023653, 0.4577766, 0.6158895)^t$

f) $(1.079675, -1.260654, 2.042489, 1.995373, 2.049536)^t$

g) $(-0.6604902, 0.03700749, -0.2493513, 0.6561139)^t$

h) $(0.8318750, 1.647766, 0.9189856, 1.791281, 0.8712129, 1.959155)^t$

2. Two iterations of the SOR method with $\omega = 1.3$ give the following results.

a) $\mathbf{x}^{(2)} = (-0.1040103, -0.1331814, 0.6774997)^t$

b) $\mathbf{x}^{(2)} = (0.957073, 0.9903875, 0.7206569)^t$

c) $\mathbf{x}^{(2)} = (2.054, 3.1005, 1.326, 1.365)^t$

d) $\mathbf{x}^{(2)} = (-0.1291297, 4.366172, -1.720288, -5.038373)^t$

e) $\mathbf{x}^{(2)} = (3.111836, -0.8661146, 0.2403225, 0.7265333)^t$

f) $\mathbf{x}^{(2)} = (0.7064258, -0.4103876, 2.417063, 2.251955, 1.061507)^t$

g) $\mathbf{x}^{(2)} = (-0.7268893, 0.1251483, -0.2923371, 0.7037018)^t$

h) $\mathbf{x}^{(2)} = (1.161875, 1.853109, 1.187423, 1.908969, 1.162713, 2.027394)^t$

3. The SOR method with $\omega = 1.2$ gives the following results.

a) $\mathbf{x}^{(12)} = (0.03488469, -0.2366474, 0.6579013)^t$

b) $\mathbf{x}^{(7)} = (0.9958341, 0.9579041, 0.7915756)^t$

c) $\mathbf{x}^{(8)} = (1.950315, 2.899950, 1.200034, 1.499996)^t$

d) $\mathbf{x}^{(8)} = (-0.08276995, 3.7896231, -1.519177, -4.777632)^t$

e) Does not converge in 25 iterations.

f) $\mathbf{x}^{(10)} = (0.7866310, -1.002807, 1.866530, 1.912645, 1.989792)^t$

g) $\mathbf{x}^{(7)} = (-0.7534489, 0.04106617, -0.2808146, 0.6918049)^t$

h) $\mathbf{x}^{(7)} = (0.9999442, 1.999934, 1.000033, 1.999958, 0.9999815, 2.000007)^t$

4. The tridiagonal matrices are in parts (b) and (d).
(5b): For $\omega = 1.012823$ we have $\mathbf{x}^{(4)} = (0.9957846, 0.9578935, 0.7915788)^t$.
(5d): For $\omega = 1.086506$ we have $\mathbf{x}^{(6)} = (-0.08277682, 3.789622, -1.519180, -4.777617)^t$.

EXERCISE SET 7.6 (*Page* 266)

1. The $\|\cdot\|_\infty$ condition numbers are:

a) 50 **b)** 241.37 **c)** 235.23 **d)** 60002

e) 339866 **f)** 12 **g)** 52 **h)** 198.17

2.

	$\|\mathbf{x} - \hat{\mathbf{x}}\|_\infty$	$K_\infty(A)\|\mathbf{b} - A\hat{\mathbf{x}}\|_\infty / \|A\|_\infty$
a)	8.571429×10^{-4}	1.238095×10^{-2}
b)	0.1	3.832060
c)	1.405510×10^{-2}	1.772950
d)	0.04	8.000000×10^{-2}
e)	20	1.152440×10^{5}
f)	0.1	3.727412×10^{-1}
g)	0.5	1.2
h)	6.551700×10^{-2}	9.059201

3. The matrix is ill-conditioned since $K_\infty = 60,000$. We have $\tilde{\mathbf{x}} = (-1.0000, 2.0000)^t$.

4. $(1.818192, 0.5909091)^t$; A is ill-conditioned since A is nearly singular.

5. **a)** We have $\tilde{\mathbf{x}} = (188.9998, 92.99998, 45.00001, 27.00001, 21.00002)^t$.

b) The condition number is $K_\infty = 80$.

c) The exact solution is $\mathbf{x} = (189, 93, 45, 27, 21)^t$.

6. **a)** $K_\infty(H^{(4)}) = 28,375$ **b)** $K_\infty(H^{(5)}) = 943,656$ **c)** $-369, 5190, -13900, 9580$

CHAPTER 8 APPROXIMATION THEORY

EXERCISE SET 8.2 (*Page* 275)

1. The linear least-squares polynomial is $P_1(x) = 1.70784x + 0.89968$.

2. $P_2(x) = 0.4066667 + 1.154848x + 0.03484848x^2$.

3. The least-squares polynomials with their errors are, respectively,
$P_1(x) = 0.6208950 + 1.219621x$, with $E = 2.719 \times 10^{-5}$;
$P_2(x) = 0.5965807 + 1.253293x - 0.01085343x^2$, with $E = 1.801 \times 10^{-5}$;
$P_3(x) = 0.6290193 + 1.185010x + 0.03533252x^2 - 0.01004723x^3$, with $E = 1.741 \times 10^{-5}$.

4. The least-squares polynomials with their errors are, respectively,
$P_1(x) = 0.9295140 + 0.5281021x$, with 2.457×10^{-2}.
$P_2(x) = 1.011341 - 0.3256988x + 1.147330x^2$, with 9.453×10^{-4}.
$P_3(x) = 1.000440 - 0.001540986x - 0.011505675x^2 + 1.021023x^3$ with 1.112×10^{-4}.
$P_4(x) = 0.9994951 + 0.1106990x - 0.8254245x^2 + 2.782884x^3 - 1.167916x^4$ with 8.608×10^{-5}.

5. **a)** The linear least-squares polynomial of degree one is $P_1(x) = 72.0845x - 194.138$ with an error of 329.

b) The least-squares polynomial of degree two is $P_2(x) = 6.61822x^2 - 1.14357x + 1.23570$ with an error of 1.44×10^{-3}.

c) The least-squares polynomial of degree three is $P_3(x) = -0.0137352x^3 + 6.84659x^2 - 2.38475x + 3.43896$ with an error of 5.27×10^{-4}.

6. **a)** The linear least-square polynomial is $P_1(x) = 1.665540x - 0.5124568$ with an error of 0.33559.

b) The least-square polynomial of degree two is $P_2(x) = 1.129424x^2 - 0.3114035x + 0.08514401$ with an error of 2.4199×10^{-3}.

c) The least-square ploynomial of degree three is $P_3(x) = 0.2662081x^3 + 0.4029322x^2 + 0.2483857x - 0.01840140$ with an error of 5.0747×10^{-6}.

7. Point average = 0.101(ACT score) + 0.487

8. $1.600393x + 25.92175$

EXERCISE SET 8.3 (*Page* 284)

1. The linear least-square approximations are:

a) $P_1(x) = 1.833333 + 4x$
b) $P_1(x) = -1.600003 + 3.600003x$
c) $P_1(x) = 1.140981 - 0.2958375x$
d) $P_1(x) = 0.1945267 + 3.000001x$
e) $P_1(x) = 0.7307083 - 0.1777249x$
f) $P_1(x) = -1.861455 + 1.666667x$

2. The least squares approximations of degree two are:

a) $P_2(x) = 2.000002 + 2.999991x + 1.000009x^2$
b) $P_2(x) = 0.4000163 - 2.400054x + 3.000028x^2$
c) $P_2(x) = 1.723551 - 0.9313682x + 0.1588827x^2$
d) $P_2(x) = 1.167179 + 0.08204442x + 1.458979x^2$
e) $P_2(x) = 0.4880050 + 0.8291881x - 0.7375172x^2$
f) $P_2(x) = -0.9089523 + 0.6275723x + 0.2597736x^2$

3. The linear least-square approximations on $[-1, 1]$ are:

a) $P_1(x) = 3.333333 - 2x$
b) $P_1(x) = 0.6000025x$
c) $P_1(x) = 0.5493063 - 0.2958375x$
d) $P_1(x) = 1.175201 + 1.103639x$
e) $P_1(x) = 0.4207355 + 0.4353975x$
f) $P_1(x) = 0.6479184 + 0.5281226x$

4. The least squares approximation of degree two on $[-1, 1]$ are:

a) $P_2(x) = 3 - 2x + 1.000009x^2$
b) $P_2(x) = 0.6000025x$
c) $P_2(x) = 0.4963454 - 0.2958375x + 0.1588827x^2$
d) $P_2(x) = 0.9962918 + 1.103639x + 0.5367282x^2$
e) $P_2(x) = 0.4982798 + 0.4353975x - 0.2326330x^2$
f) $P_2(x) = 0.6947898 + 0.5281226x - 0.1406141x^2$

5. The errors for the approximations in Exercise 3 are:

a) 0.177779
b) 0.0457206
c) 0.00484624
d) 0.0526541
e) 0.0153784
f) 0.00363453

6. The errors for the approximations in Exercise 4 are:

a) 0
b) 0.0457206
c) 0.00035851
d) 0.0014082
e) 0.00575753
f) 0.00011949

7. The Gram-Schmidt process produces the following collections of polynomials:

a) $\phi_0(x) = 1, \phi_1(x) = x - 0.5, \phi_2(x) = x^2 - x + \frac{1}{6}$, and $\phi_3(x) = x^3 - 1.5x^2 + 0.6x - 0.05$.

b) $\phi_0(x) = 1, \phi_1(x) = x - 1$, $\phi_2(x) = x^2 - 2x + \frac{2}{3}$, and $\phi_3(x) = x^3 - 3x^2 + \frac{12}{5}x - \frac{2}{5}$.

c) $\phi_0(x) = 1, \phi_1(x) = x - 2, \phi_2(x) = x^2 - 4x + \frac{11}{3}$, and $\phi_3(x) = x^3 - 6x^2 + 11.4x - 6.8$.

8. The Gram-Schmidt process produces the following collections of polynomials:

a) $3.833333\phi_0(x) + 4.000000\phi_1(x)$

b) $2\phi_0(x) + 3.6\phi_1(x)$

c) $0.5493061\phi_0(x) - 0.2958369\phi_1(x)$

d) $3.194528\phi_0(x) + 3\phi_1(x)$

e) $0.6567600\phi_0(x) + 0.09167105\phi_1(x)$

f) $1.471878\phi_0(x) + 1.666667\phi_1(x)$

9. The least-square polynomials of degree two are:

a) $P_2(x) = 3.83333\phi_0(x) + 4\phi_1(x) + 0.999999\phi_2(x)$

b) $P_2(x) = 2\phi_0(x) + 3.6\phi_1(x) + 3\phi_2(x)$

c) $P_2(x) = 0.549306\phi_0(x) - 0.295837\phi_1(x) + 0.158878\phi_2(x)$

d) $P_2(x) = 3.194528\phi_0(x) + 3\phi_1(x) + 1.458960\phi_2(x)$

e) $P_2(x) = 0.0656760\phi_0(x) + 0.0916711\phi_1(x) - 0.737512\phi_2(x)$

f) $P_2(x) = 1.47188\phi_0(x) + 1.66667\phi_1(x)^3 + 0.259771\phi_2(x)$

10. The least-square polynomials of degree three are:

a) $P_3(x) = 3.833333\phi_0(x) + 4.000000\phi_1(x) + 0.9999998\phi_2(x)$

b) $P_3(x) = 2\phi_0(x) + 3.6\phi_1(x) + 3\phi_2(x) + \phi_3(x)$

c) $P_3(x) = 0.5493061\phi_0(x) - 0.2958369\phi_1(x) + 0.1588784\phi_2(x)$

d) $P_3(x) = 3.194528\phi_0(x) + 3\phi_1(x) + 1.458960\phi_2(x) + 0.4787959\phi_3(x)$

e) $P_3(x) = 0.6567600\phi_0(x) + 0.09167105\phi_1(x) - 0.7375118\phi_2(x)$

f) $P_3(x) = 1.471878\phi_0(x) + 1.666667\phi_1(x) + 0.2597705\phi_2(x)$

11. The Laguerre polynomials are $L_1(x) = x - 1$, $L_2(x) = x^2 - 4x + 2$, and $L_3(x) = x^3 - 9x^2 + 18x - 6$.

12. **a)** $2L_0(x) + 4L_1(x) + L_2(x)$

b) $\frac{1}{2}L_0(x) - \frac{1}{4}L_1(x) + \frac{1}{16}L_2(x) + \frac{1}{96}L_3(x)$

c) $6L_0(x) + 18L_1(x) + 9L_2(x) - L_3(x)$

d) $\frac{1}{3}L_0(x) - \frac{2}{9}L_1(x) + \frac{2}{27}L_2(x) + \frac{4}{243}L_3(x)$

EXERCISE SET 8.4 (*Page* 290)

1. The interpolating polynomials of degree two are given below:

a) $P_2(x) = 2.377443 + 1.590534(x - 0.8660254) + 0.5320418(x - 0.8660254)x$

b) $P_2(x) = 0.7617600 + 0.8796047(x - 0.8660254)$

c) $P_2(x) = 1.052926 + 0.4154370(x - 0.8660254) - 0.1384262x(x - 0.8660254)$

d) $P_2(x) = 0.5625 + 0.649519(x - 0.8660254) + 0.75x(x - 0.8660254)$

2. Bounds for the maximum errors of polynomials in Exercise 1 are:

a) 0.1132617 **b)** 0.04166667 **c)** 0.08333333 **d)** 1.000000

3. The interpolating polynomials of degree three are given below:

a) $P_3(x) = 2.519044 + 1.945377(x - 0.9238795)$
$+0.7047420(x - 0.9238795)(x - 0.3826834)$
$+0.1751757(x - 0.9238795)(x - 0.3826834)(x + 0.3826834)$

b) $P_3(x) = 0.7979459 + 0.7844380(x - 0.9238795)$
$-0.1464394(x - 0.9238795)(x - 0.3826834)$
$-0.1585049(x - 0.9238795)(x - 0.3826834)(x + 0.3826834)$

c) $P_3(x) = 1.072911 + 0.3782067(x - 0.9238795)$
$-0.09799213(x - 0.9238795)(x - 0.3826834)$
$+0.04909073(x - 0.9238795)(x - 0.3826834)(x + 0.3826834)$

d) $P_3(x) = 0.7285533 + 1.306563(x - 0.9238795)$
$+0.9999999(x - 0.9238795)(x - 0.3826834)$

4. Bounds for the maximum errors of polynomials in Exercise 3 are:

a) 0.01415772 **b)** 0.004382661 **c)** 0.03125000 **d)** 0.1250000

5. The zeros of $\tilde{T}_3$ produce the following interpolating polynomials of degree two.

a) $P_2(x) = 0.3489153 - 0.1744576(x - 2.866025) + 0.1538462(x - 2.866025)(x - 2)$

b) $P_2(x) = 0.1547375 - 0.2461152(x - 1.866025) + 0.1957273(x - 1.866025)(x - 1)$

c) $P_2(x) = 0.6166200 - 0.2370869(x - 0.9330127)$
$-0.7427732(x - 0.9330127)(x - 0.5)$

d) $P_2(x) = 3.0177125 + 1.883800(x - 2.866025) + 0.2584625(x - 2.866025)(x - 2)$

6. The zeros of $\tilde{T}_4$ produce the following interpolating polynomials of degree three.

a) $P_3(x) = 0.3420114 - 0.1435404(x - 2.9238795) + 0.08875220(x - 2.9238795)(x - 2.3826834)$
$-0.08247422(x - 2.9238795)(x - 2.3826834)(x - 1.6173166)$

b) $P_3(x) = 0.1460393 - 0.1937654(x-1.9238795) + 0.1401840(x-1.9238795)(x-1.3826834)$
$-0.06444354(x-1.9238795)(x-1.3826834)(x-0.6173166)$

c) $P_3(x) = 0.5987353 - 0.4209688(x-0.9619398) - 0.8177026(x-0.9619398)(x-0.6913417)$
$-0.1889195(x-0.9619398)(x-0.6913417)(x-0.3086583)$

d) $P_3(x) = 3.137063 + 1.974058(x-2.9238795) + 0.2197224(x-2.9238795)(x-2.3826834)$
$-0.04516458(x-2.9238795)(x-2.3826834)(x-1.6173166)$

7. If $i > j$, then

$$\frac{1}{2}(T_{i+j}(x) + T_{i-j}(x)) = \frac{1}{2}\Big(\cos(i+j)\theta + \cos(i-j)\theta\Big) = \cos i\theta \cos j\theta = T_i(x)T_j(x).$$

8. The change of variable $x = \cos\theta$ produces

$$\int_{-1}^{1} \frac{T_n^2(x)}{\sqrt{1-x^2}}\,dx = \int_{-1}^{1} \frac{[\cos(n \arccos x)]^2}{\sqrt{1-x^2}}\,dx = \int_0^{\pi} \cos^2(n\theta)\,d\theta = \frac{\pi}{2}.$$

EXERCISE SET 8.5 (Page 297)

1. The Padé approximations of degree two for $f(x) = e^{2x}$ are:
$n = 2, m = 0 : r_{2,0}(x) = 1 + 2x + 2x^2$
$n = 1, m = 1 : r_{1,1}(x) = (1+x)/(1-x)$
$n = 0, m = 2 : r_{0,2}(x) = (1 - 2x + 2x^2)^{-1}$

i	x_i	$f(x_i)$	$r_{2,0}(x_i)$	$r_{1,1}(x_i)$	$r_{0,2}(x_i)$
1	0.2	1.4918	1.4800	1.5000	1.4706
2	0.4	2.2255	2.1200	2.3333	1.9231
3	0.6	3.3201	2.9200	4.0000	1.9231
4	0.8	4.9530	3.8800	9.0000	1.4706
5	1.0	7.3891	5.0000	undefined	1.0000

2. The Padé approximations of degree three for $f(x) = x\ln(x+1)$ are:
$m = 0, n = 3 : x^2 - \frac{1}{2}x^3$
$m = 1, n = 2 : \frac{x^2}{1+\frac{1}{2}x}$
no others

3. The Padé approximation for $f(x) = e^x$ with $n = 2$ and $m = 3$ is
$r_{2,3}(x) = (1 + \frac{2}{5}x + \frac{1}{20}x^2)/(1 - \frac{3}{5}x + \frac{3}{20}x^2 - \frac{1}{60}x^3)$

4. $r_{3,2}(x) = (1 + \frac{3}{5}x + \frac{3}{20}x^2 + \frac{1}{60}x^3)/(1 - \frac{2}{5}x + \frac{1}{20}x^2)$

5. $r_{3,3}(x) = (x - \frac{7}{60}x^3)/(1 + \frac{1}{20}x^2)$

6. **a)** $r_{2,4}(x) = x/(1 + \frac{1}{6}x^2 + \frac{7}{360}x^4)$

b) $r_{4,2}(x) = (x - \frac{7}{60}x^3)/(1 + \frac{1}{20}x^2)$

7. The Padé approximations of degree five are given below:

a) $r_{0,5}(x) = (1 + x + \frac{1}{2}x^2 + \frac{1}{6}x^3 + \frac{1}{24}x^4 + \frac{1}{120}x^5)^{-1}$

b) $r_{1,4}(x) = (1 - \frac{1}{5}x)/(1 + \frac{4}{5}x + \frac{3}{10}x^2 + \frac{1}{15}x^3 + \frac{1}{120}x^4)$

c) $r_{3,2}(x) = (1 - \frac{3}{5}x + \frac{3}{20}x^2 - \frac{1}{60}x^3)/(1 + \frac{2}{5}x + \frac{1}{20}x^2)$

d) $r_{4,1}(x) = (1 - \frac{4}{5}x + \frac{3}{10}x^2 - \frac{1}{15}x^3 + \frac{1}{120}x^4)/(1 + \frac{1}{5}x)$

8. **a)**

$$1 + \cfrac{4}{x - \frac{5}{4} + \cfrac{\frac{21}{16}}{x + \frac{1}{4}}}$$

b)

$$\cfrac{4}{2x - \cfrac{1}{2} + \cfrac{\frac{23}{8}}{x - \cfrac{\frac{63}{92}}{1 - \cfrac{\frac{232}{207}}{x + \cfrac{23}{9}}}}}$$

c)

$$2x - 7 + \cfrac{10}{x - \cfrac{3}{10} + \cfrac{\frac{469}{100}}{x - \cfrac{23}{10}}}$$

d)

$$\cfrac{2}{3 + \cfrac{0.5}{x - 0.5 + \cfrac{7}{x + 1.2857142 - \cfrac{6.6326530}{x - 0.28571428}}}}$$

9. For

8(a):	**a)** 5.63	**b)** 5.63	**c)** 5.62, exact value 5.61.
8(b):	**a)** 0.303	**b)** 0.304	**c)** 0.303, exact value 0.304.
8(c):	**a)** −0.112	**b)** −0.112	**c)** −0.120, exact value −0.113.
8(d):	**a)** 0.836	**b)** 0.837	**c)** 0.836, exact value 0.836.

10. **a)** $e^{-x} \approx 1.266066T_0 - 1.130318T_1$ **b)** $\cos x \approx 0.7651975T_0 - 0.2298070T_2$

c) $\sin x \approx 0.8800998T_1 - 0.03912761T_3 + 0.0004995155T_5 - 0.000003004652T_7$

d) $e^{x} \approx 1.266066T_0 + 1.130318T_1 + 0.2714953T_2 + 0.04433685T_3 + 0.005474240T_4$

11. $r_{T_{2,0}}(x) = (1.266066T_0(x) - 1.130318T_1(x) + 0.2714953T_2(x))/T_0(x)$
$r_{T_{1,1}}(x) = (0.9945705T_0(x) - 0.4569046T_1(x))/(T_0(x) + 0.48038745T_1(x))$,
$r_{T_{0,2}}(x) = 0.7940220T_0(x)/(T_0(x) + 0.8778575T_1(x) + 0.1774266T_2(x))$.

x	$f(x)$	$r_{T_{2,0}}$	$r_{T_{1,1}}$	$r_{T_{0,2}}$
0.25	0.77801	0.745928	0.785954	0.746110
0.5	0.606531	0.565159	0.617741	0.588071
1.0	0.367879	0.407243	0.363193	0.386332

12. $m = 3, n = 0$ and $m = 2, n = 1 : \frac{0.7306893T_0}{T_0+0.3003238T_2}$
$m = 1, n = 2$ and $m = 0, n = 3 : \frac{0.7651975T_0-0.2298070T_2}{T_0}$

13. $r_{T_{2,2}}(x) = \dfrac{0.91747T_1(x)}{T_0(x) + 0.088863T_2(x)}$

14. $m = 5, n = 0 : \frac{0.7898486T_0}{-0.8927799T_1+0.2144414T_2-0.03502476T_3+0.004335741T_4-0.0004335974T_5}$

$m = 4, n = 1 : \frac{0.8698859T_0+0.1792990T_1}{T_0-0.7319036T_1+0.1308634T_2-0.01374200T_3+0.0007516311T_4}$

$m = 3, n = 2 : \frac{0.9455983T_0+0.3537814T_1+0.02028345T_2}{T_0-0.5848039T_1+0.07467597T_2-0.004402997T_3}$

$m = 2, n = 3 : \frac{1.055167T_0+0.6127802T_1+0.07740801T_2+0.004495996T_3}{T_0-0.3785111T_1+0.02224353T_2}$

$m = 1, n = 4 : \frac{1.153963T_0+0.8522588T_1+0.1549949T_2+0.01686746T_3+0.001023136T_4}{T_0-0.1983568T_1}$

$m = 0, n = 5 : 1.266066T_0 + 1.130318T_1 + 0.2714953T_2 + 0.04433685T_3$
$+0.005474240T_4 + 0.0005429263T_5$

EXERCISE SET 8.6 (*Page* 303)

1. $S_2(x) = 2\sin x$

2. $S_n(x) = 2\sum_{k=1}^{n-1} \frac{(-1)^{k+1}}{k} \sin kx$

3. $S_2(x) = 9.214561 - 6.515678\cos x + 2.606271\cos 2x + 6.515678\sin x$

4. The general trigonometric least-squares polynomial is $S_n(x) = \sum_{k=1}^{n-1} \frac{2}{k\sqrt{\pi}}(1-(-1)^k)\sin kx$.

5. The trigonometric least-squares polynomials are given below.

a) $S_2(x) = \cos 2x$ **b)** $S_2(x) = 0$

c) $S_3(x) = 3.132905 + 0.5886815 \cos x - 0.2700642 \cos 2x + 0.2175679 \cos 3x + 0.8341640 \sin x - 0.3097866 \sin 2x$

d) $S_3(x) = -4.092652 + 3.883872 \cos x - 2.320482 \cos 2x + 0.7310818 \cos 3x$

6. **a)** $E(S_2) = 0$ **b)** $E(S_2) = 4$

c) $E(S_3) = 0.8259814$ **d)** $E(S_3) = 1.936668$

7. The trigonometric least-squares polynomial is $S_3(x) = -0.9937858 + 0.2391965 \cos x + 1.515393 \cos 2x + 0.2391965 \cos 3x - 1.150649 \sin x$ with error $E(S_3) = 7.271197$.

8. The trigonometric least-squares polynomial is

$$S_3(x) = 0.1240293 - 0.8600803 \cos x + 2.549330 \cos 2x - 0.6409933 \cos 3x - 0.8321197 \sin x - 0.6695062 \sin 2x$$

with error 107.913.
The approximation in Exercise 8 is better because, in this case, $\sum_{j=0}^{n} (f(\xi_j) - S_3(\xi_j))^2 = 397.3678$, whereas the approximation in Exercise 7 has $\sum_{j=0}^{n} (f(\xi_j) - S_3(\xi_j))^2 = 569.3589$.

9. The trigonometric least-squares polynomials and their errors are given below.

a) $S_3(x) = -0.08676065 - 1.446416 \cos \pi(x-3) - 1.617554 \cos 2\pi(x-3) + 3.980729 \cos 3\pi(x-3) - 2.154320 \sin \pi(x-3) + 3.907451 \sin 2\pi(x-3)$ with $E(S_3) = 210.90453$.

b) $S_3(x) = -0.0867607 - 1.446416 \cos \pi(x-3) - 1.617554 \cos 2\pi(x-3) + 3.980729 \cos 3\pi(x-3) - 2.354008 \cos 4\pi(x-3) - 2.154320 \sin \pi(x-3) + 3.907451 \sin 2\pi(x-3) - 1.166181 \sin 3\pi(x-3)$ with $E(S_4) = 169.4943$.

10. **a)** The trigonometric least-squares polynomial is $S_4(x) = 0.4205545 - 0.09802618 \cos x + 0.002905236 \cos 2x + 0.01328013 \cos 3x - 0.01884123 \cos 4x + 0.2398368 \sin x - 0.1290849 \sin 2x + 0.08577849 \sin 3x$.

b) $\int_0^1 S_4(x)dx = 0.2102773$ **c)** $\int_0^1 x^2 \sin x dx = 0.2232443$

EXERCISE SET 8.7 (*Page* 310)

1. The trigonometric interpolating polynomials are:

a) $S_2(x) = -12.33701 + 4.934802 \cos x - 2.467401 \cos 2x + 4.934802 \sin x$

b) $S_2(x) = -6.16851 + 9.869604 \cos x - 3.701102 \cos 2x + 4.934802 \sin x$

c) $S_2(x) = 1.570796 - 1.570796 \cos x$

d) $S_2(x) = -0.5 - 0.5 \cos 2x + \sin x$

2. The trigonometric interpolating polynomial is $S_4(x) = -4.626377 + 6.679518 \cos x - 3.701102 \cos 2x + 3.190086 \cos 3x - 1.542126 \cos 4x + 5.956833 \sin x - 2.467401 \sin 2x + 1.022031 \sin 3x$

3. The Fast Fourier Transform method gives the trigonometric interpolating polynomials listed below.
   a) $S_4(x) = -11.10331 + 2.467401 \cos x - 2.467401 \cos 2x + 2.467401 \cos 3x - 1.233701 \cos 4x + 5.956833 \sin x - 2.467401 \sin 2x + 1.022030 \sin 3x$
   b) $S_4(x) = 1.570796 - 1.340756 \cos x - 0.2300378 \cos 3x$
   c) $S_4(x) = -0.1264264 + 0.2602724 \cos x - 0.3011140 \cos 2x + 1.121372 \cos 3x + 0.04589648 \cos 4x - 0.1022190 \sin x + 0.2754062 \sin 2x - 2.052955 \sin 3x$
   d) $S_4(x) = -0.1526819 + 0.04754278 \cos x + 0.6862114 \cos 2x - 1.216913 \cos 3x + 1.176143 \cos 4x - 0.8179387 \sin x + 0.1802450 \sin 2x + 0.2753402 \sin 3x$

4. a) The trigonometric interpolating polynomial is $S_4(x) = 0.3471000 - 0.02475498 \cos x - 0.0697570 \cos 2x + 0.08468317 \cos 3x - 0.04386479 \cos 4x + 0.2268260 \sin x - 0.1021640 \sin 2x + 0.04284648 \sin 3x$
   b) 0.1735500
   c) 0.2232443

5.

	Approximation	Actual
a)	−69.76412	−62.01255
b)	9.869605	9.869604
c)	−0.7943605	−0.2739384
d)	−0.9593284	−0.9570636

6. The b_k terms are all zero. The a_k terms are $a_0 = -4.01287586$, $a_0 = 3.80276903$, $a_0 = -2.23519870$, $a_0 = 0.63810403$, $a_0 = -0.31550821$, $a_0 = 0.19408145$, $a_0 = -0.13464491$, $a_0 = 0.10100593$, $a_0 = -0.08015708$, $a_0 = 0.06643598$, $a_0 = -0.05704353$, $a_0 = 0.05046675$, $a_0 = -0.04583431$, $a_0 = 0.04262318$, $a_0 = -0.04051395$, $a_0 = 0.03931584$, $a_0 = -0.03892713$.

CHAPTER 9 APPROXIMATING EIGENVALUES

EXERCISE SET 9.2 (*Page* 317)

1. **a)** The eigenvalues and associated eigenvectors are $\lambda_1 = 2$, $\mathbf{v}^{(1)} = (1,0,0)^t$; $\lambda_2 = 1$, $\mathbf{v}^{(2)} = (0,2,1)^t$; and $\lambda_3 = -1$, $\mathbf{v}^{(3)} = (-1,1,1)^t$. Yes, the set is linearly independent.

b) The eigenvalues and associated eigenvectors are $\lambda_1 = \lambda_2 = 2$, $\mathbf{v}^{(1)} = \mathbf{v}^{(2)} = (1,0,0)^t$; and $\lambda_3 = 3$, $\mathbf{v}^{(3)} = (0,1,1)^t$. No.

c) The eigenvalues and associated eigenvectors are $\lambda_1 = \lambda_2 = \lambda_3 = 2$, $\mathbf{v}^{(1)} = \mathbf{v}^{(2)} = (1,0,0)^t$; and $\mathbf{v}^{(3)} = (0,1,0)$. No.

d) The eigenvalues and associated eigenvectors are $\lambda_1 = \lambda_2 = \lambda_3 = 1$, $\mathbf{v}^{(1)} = \mathbf{v}^{(2)} = (1,0,1)^t$ and $\mathbf{v}^{(3)} = (0,1,1)$. No.

e) The eigenvalues and associated eigenvectors are $\lambda_1 = 2$, $\mathbf{v}^{(1)} = (0,1,0)^t$; $\lambda_2 = 3$, $\mathbf{v}^{(2)} = (1,0,1)^t$; and $\lambda_3 = 1$, $\mathbf{v}^{(3)} = (1,0,-1)^t$. Yes, the set is linearly independent.

f) The eigenvalues and associated eigenvectors are $\lambda_1 = \lambda_2 = 3$, $\mathbf{v}^{(1)} = (1,0,-1)^t, \mathbf{v}^{(2)} = (0,1,-1)^t$; and $\lambda_3 = 0$, $\mathbf{v}^{(3)} = (1,1,1)^t$. Yes, the set is linearly independent.

g) The eigenvalues and associated eigenvectors are $\lambda_1 = 1, \mathbf{v}^{(1)} = (1,0,-1)^t; \lambda_2 = 1+\sqrt{2}, \mathbf{v}^{(2)} = (\sqrt{2},1,1)^t$; and $\lambda_3 = 1-\sqrt{2}, \mathbf{v}^{(3)} = (-\sqrt{2},1,1)^t$; Yes, the set is linearly independent.

h) The eigenvalues and associated eigenvectors are $\lambda_1 = 1$, $\mathbf{v}^{(1)} = (1,0,-1)^t; \lambda_2 = 1, \mathbf{v}^{(2)} = (1,-1,0)^t$; and $\lambda_3 = 4$, $\mathbf{v}^{(4)} = (1,1,1)^t$. Yes, the set is linearly independent.

2. **a)** The matricies in 1(e) and 1(h) are positive definite.

b) $1(\text{e})$: $P = \begin{bmatrix} 0 & 1 & 1 \\ 1 & 0 & 0 \\ 0 & 1 & -1 \end{bmatrix}$, $\quad 1(\text{h})$: $P = \begin{bmatrix} 1 & 1 & 1 \\ 0 & -2 & 1 \\ -1 & 1 & 1 \end{bmatrix}$

3. **a)** The three eigenvalues are within $\{\lambda \mid |\lambda| \le 2\} \cup \{\lambda \mid |\lambda - 2| \le 2\}$.

b) The three eigenvalues are within $R_1 = \{\lambda \mid |\lambda - 4| \le 2\}$.

c) The three real eigenvalues satisfy $0 \le \lambda \le 6$.

d) The three real eigenvalues satisfy $1.25 \le \lambda \le 8.25$.

e) The four real eigenvalues satisfy $-4 \le \lambda \le 1$.

f) The four real eigenvalues are within $R_1 = \{\lambda \mid |\lambda - 2| \le 4\}$.

4. Let $\mathbf{w} = (w_1, w_2, w_3)^t, \mathbf{x} = (x_1, x_2, x_3)^t, \mathbf{y} = (y_1, y_2, y_3)^t$ and $\mathbf{z} = (z_1, z_2, z_3)^t$ be in $\mathbb{R}^3$. Suppose $\mathbf{w}, \mathbf{x}$, and $\mathbf{y}$ are linearly independent. Consider the linear system

$$\begin{bmatrix} w_1 & x_1 & y_1 \\ w_2 & x_2 & y_2 \\ w_3 & x_3 & y_3 \end{bmatrix} \begin{bmatrix} a \\ b \\ c \end{bmatrix} = \begin{bmatrix} 0 \\ 0 \\ 0 \end{bmatrix}.$$

Since $\mathbf{w}, \mathbf{x}$, and $\mathbf{y}$ are linearly independent, the only solution is $a = b = c = 0$. Thus, the determinant of the matrix is nonzero. Hence, the linear system

$$\begin{bmatrix} w_1 & x_1 & y_1 \\ w_2 & x_2 & y_2 \\ w_3 & x_3 & y_3 \end{bmatrix} \begin{bmatrix} a \\ b \\ c \end{bmatrix} = \begin{bmatrix} z_1 \\ z_2 \\ z_3 \end{bmatrix}$$

has a unique solution. Thus, $a\mathbf{w} + b\mathbf{x} + c\mathbf{y} - \mathbf{z} = \mathbf{0}$, so $\{\mathbf{w}, \mathbf{x}, \mathbf{y}, \mathbf{z}\}$ is linearly dependent. If $\{\mathbf{w}, \mathbf{x}, \mathbf{y}\}$ is linearly dependent, then $\{\mathbf{w}, \mathbf{x}, \mathbf{y}, \mathbf{z}\}$ also is linearly dependent.

5. If $c_1\mathbf{v}_1 + \cdots + c_k\mathbf{v}_k = \mathbf{0}$, then for any j, with $1 \le j \le k$, we have $c_1\mathbf{v}_j^t\mathbf{v}_1 + \cdots + c_k\mathbf{v}_j^t\mathbf{v}_k = 0$. But orthogonality gives $c_j\mathbf{v}_j^t\mathbf{v}_j = 0$, so $c_j = 0$.

6. Since $\{\mathbf{v}_i\}_{i=1}^n$ is linear independent in $\mathbb{R}^n$, there exist numbers $c_1, \ldots, c_n$ with

$$\mathbf{x} = c_1\mathbf{v}_1 + \cdots + c_n\mathbf{v}_n.$$

Hence, for any j,with $1 \le j \le n$,

$$\mathbf{v}_j^t\mathbf{x} = c_1\mathbf{v}_j^t\mathbf{v}_1 + \cdots + c_n\mathbf{v}_j^t\mathbf{v}_n = c_j\mathbf{v}_j^t\mathbf{v}_j = c_j.$$

7. Let $A\mathbf{x}^{(i)} = \lambda_i\mathbf{x}^{(i)}$ for $i = 1, 2, \ldots, n$ where the λ_i are distinct. Suppose $\{\mathbf{x}^{(i)}\}_{i=1}^k$ is the largest linearly independent set of eigenvectors of A where $1 \le k < n$. (Note that a re-indexing may be necessary for the preceding statement to hold.)
Since $\{\mathbf{x}^{(i)}\}_{i=1}^{k+1}$ is linearly dependent, there exist numbers $c_1, \ldots, c_{k+1}$, not all zero, with

$$c_1\mathbf{x}^{(1)} + \ldots + c_k\mathbf{x}^{(k)} + c_{k+1}\mathbf{x}^{(k+1)} = \mathbf{0}.$$

Since $\{\mathbf{x}^{(i)}\}_{i=1}^k$ is linearly independent, $c_{k+1} \ne 0$. Multiplying by A gives

$$c_1\lambda_1\mathbf{x}^{(1)} + \ldots + c_k\lambda_k\mathbf{x}^{(k)} + c_{k+1}\lambda_{k+1}\mathbf{x}^{(k+1)} = \mathbf{0}.$$

Thus,

$$\frac{c_1(\lambda_{k+1}-\lambda_1)}{c_{k+1}}\mathbf{x}^{(1)}+\ldots+\frac{c_k(\lambda_{k+1}-\lambda_k)}{c_{k+1}}\mathbf{x}^{(k)}=\mathbf{0}.$$

But $\{\mathbf{x}^{(i)}\}_{i=1}^{k}$ is linearly independent and $\mathbf{x}^{(k+1)}\neq\mathbf{0}$, so $\lambda_{k+1}=\lambda_i$ for some $1\leq i\leq k$.

8. **a)** The eigenvalues are $\lambda_1 = 5.307857563$, $\lambda_2 = -0.4213112993$, $\lambda_3 = -0.1365462647$ with associated eigenvectors $(0.59020967, 0.51643129, 0.62044441)^t$, $(0.77264234, -0.13876278, -0.61949069)^t$, and $(0.23382978, -0.84501102, 0.48091581)^t$, respectively.

b) A is not positive definite, since $\lambda_2 < 0$ and $\lambda_3 < 0$.

EXERCISE SET 9.3 (*Page* 327)

1. The approximate eigenvalues and approximate eigenvectors are:

a) $\mu^{(3)} = 3.666667$, $\mathbf{x}^{(3)} = (0.9772727, 0.9318182, 1)^t$

b) $\mu^{(3)} = 2.000000$, $\mathbf{x}^{(3)} = (1, 1, 0.5)^t$

c) $\lambda = \mu^{(2)} = 5$, $\mathbf{x} = \mathbf{x}^{(2)} = (0.25, 1, 0.25)^t$

d) $\mu^{(3)} = 5.000000$, $\mathbf{x}^{(3)} = (-0.2578947, 1, -0.2842105)^t$

e) $\mu^{(3)} = 5.038462$, $\mathbf{x}^{(3)} = (1, 0.2213741, 0.3893130, 0.4045802)^t$

f) $\mu^{(3)} = 7.531073$, $\mathbf{x}^{(3)} = (0.6886722, -0.6706677, -0.9219805, 1)^t$

g) $\mu^{(3)} = 8.644444$, $\mathbf{x}^{(3)} = (0.2544987, 1, -0.2519280, -0.3161954)^t$

h) $\mu^{(3)} = -3.691176$, $\mathbf{x}^{(3)} = (1, -0.4462151, -0.07968127, -0.5816733)^t$

2. The approximate eigenvalues and approximate eigenvectors are:

a) $\mu^{(3)} = 1.027730$, $\mathbf{x}^{(3)} = (-0.1889082, 1, -0.7833622)^t$

b) $\mu^{(3)} = -0.4166667$, $\mathbf{x}^{(3)} = (1, -0.75, -0.6666667)^t$

c) $\mu^{(3)} = 4.970874$, $\mathbf{x}^{(3)} = (-0.2427185, 1, -0.2427185)^t$

d) $\mu^{(3)} = 17.64493$, $\mathbf{x}^{(3)} = (-0.3805794, -0.09079132, 1)^t$

e) $\mu^{(3)} = 1.378684$, $\mathbf{x}^{(3)} = (-0.3690277, -0.2522880, 0.2077438, 1)^t$

f) $\mu^{(3)} = 3.996073$, $\mathbf{x}^{(3)} = (0.9991429, 0.9932014, 1, 0.9939825)^t$

g) $\mu^{(3)} = 3.973087$, $\mathbf{x}^{(3)} = (1, -0.5084634, -0.9850723, -0.4885946)^t$

h) $\mu^{(3)} = -2.094588$, $\mathbf{x}^{(3)} = (-0.1597981, 1, -0.5975984, -0.02937691)^t$

3. The approximate eigenvalues and approximate eigenvectors are:

a) $\mu^{(3)} = 3.959538$, $\mathbf{x}^{(3)} = (0.5816124, 0.5545606, 0.5951383)^t$

b) $\mu^{(3)} = 2.0000000$, $\mathbf{x}^{(3)} = (-0.6666667, -0.6666667, -0.3333333)^t$

c) $\mu^{(3)} = 3.142857$, $\mathbf{x}^{(3)} = (0.6767155, -0.6767155, 0.2900210)^t$

d) $\mu^{(3)} = 7.189567$, $\mathbf{x}^{(3)} = (0.5995308, 0.7367472, 0.3126762)^t$

e) $\mu^{(3)} = 6.037037$, $\mathbf{x}^{(3)} = (0.5073714, 0.4878571, -0.6634857, -0.2536857)^t$

f) $\mu^{(3)} = 5.142562$, $\mathbf{x}^{(3)} = (0.8373051, 0.3701770, 0.1939022, 0.3525495)^t$

g) $\mu^{(3)} = 8.086569$, $\mathbf{x}^{(3)} = (0.2296403, 0.9023239, -0.2273207, -0.2853107)^t$

h) $\mu^{(3)} = 8.593142$, $\mathbf{x}^{(3)} = (-0.4134762, 0.4026664, 0.5535536, -0.6003962)^t$

4. The approximate eigenvalues and approximate eigenvectors are:

a) $\lambda_1 \approx \mu^{(9)} = 3.999908$, $\mathbf{x}^{(9)} = (0.9999943, 0.9999828, 1)^t$
$\lambda_2 \approx \mu^{(1)} = 1.000000$, $\mathbf{x}^{(1)} = (-2.999908, 2.999908, 0)^t$

b) $\lambda_1 \approx \mu^{(13)} = 2.414214$, $\mathbf{x}^{(13)} = (1, 0.7071429, 0.7070707)^t$
$\lambda_2 \approx \mu^{(1)} = 1.000000$, $\mathbf{x}^{(1)} = (0, -1.414214, 1.414214)^t$

c) $\lambda_1 \approx \mu^{(2)} = 5.000000$, $\mathbf{x}^{(2)} = (-0.25, 1, -0.25)^t$
$\lambda_2 \approx \mu^{(1)} = 1.000000$, $\mathbf{x}^{(1)} = (-4, 0, 4)^t$

d) $\lambda_1 \approx \mu^{(9)} = 5.124749$, $\mathbf{x}^{(9)} = (-0.2424476, 1, -0.3199733)^t$
$\lambda_2 \approx \mu^{(6)} = 1.636734$, $\mathbf{x}^{(6)} = (1.783218, -1.135350, -3.124733)^t$

e) $\lambda_1 \approx \mu^{(24)} = 5.235861$, $\mathbf{x}^{(24)} = (1, 0.6178361, 0.1181667, 0.4999220)^t$
$\lambda_2 \approx \mu^{(12)} = 3.618177$, $\mathbf{x}^{(12)} = (0.7235746, -1.170633, 1.170688, -0.2762855)^t$

f) $\lambda_1 \approx \mu^{(17)} = 8.999667$, $\mathbf{x}^{(17)} = (0.9999085, -0.9999078, -0.9999993, 1)^t$
$\lambda_2 \approx \mu^{(21)} = 5.000051$, $\mathbf{x}^{(21)} = (1.999338, -1.999603, 1.999603, -2.000198)^t$

g) The method did not converge in 25 iterations.

h) The method did not converge in 25 iterations.

5. The approximate eigenvalues and approximate eigenvectors are:

a) $\mu^{(8)} = 4.000001$, $\mathbf{x}^{(8)} = (0.9999773, 0.99993134, 1)^t$

b) The method did not converge in 25 iterations.

c) $\mu^{(3)} = 5.000000$, $\mathbf{x}^{(3)} = (-0.25, 1, -0.25)^t$

d) $\mu^{(7)} = 5.124890$, $\mathbf{x}^{(7)} = (-0.2425938, 1, -0.3196351)^t$

e) $\mu^{(15)} = 5.236112$, $\mathbf{x}^{(15)} = (1, 0.6125369, 0.1217216, 0.4978318)^t$

f) $\mu^{(10)} = 8.999890$, $\mathbf{x}^{(10)} = (0.9944137, -0.9942148, -0.9997991, 1)^t$

g) The method did not converge in 25 iterations.

h) The method did not converge in 25 iterations.

6. The approximate eigenvalues and approximate eigenvectors are:

a) $\mu^{(8)} = 4.0000000$, $\mathbf{x}^{(8)} = (0.5773547, 0.5773282, 0.5773679)^t$

b) $\mu^{(13)} = 2.414214$, $\mathbf{x}^{(13)} = (-0.7071068, -0.5000255, -0.4999745)^t$

c) $\mu^{(18)} = 3.414214$, $\mathbf{x}^{(18)} = (0.5000660, -0.7071068, 0.4999340)^t$

d) $\mu^{(16)} = 7.223663$, $\mathbf{x}^{(16)} = (0.6247845, 0.7204271, 0.3010466)^t$

e) $\mu^{(20)} = 7.086130$, $\mathbf{x}^{(20)} = (0.3325999, 0.2671862, -0.7590108, -0.4918246)^t$

f) $\mu^{(21)} = 5.236068$, $\mathbf{x}^{(21)} = (0.7795539, 0.4815996, 0.09214214, 0.3897016)^t$

g) The method did not converge in 25 iterations.

h) $\mu^{(16)} = 9.0000000$, $\mathbf{x}^{(16)} = (-0.4999592, 0.4999584, 0.5000408, -0.5000416)^t$

7. The approximate eigenvalues and approximate eigenvectors are:

a) $\mu^{(9)} = 1.000015$, $\mathbf{x}^{(9)} = (-0.1999939, 1, -0.7999909)^t$

b) $\mu^{(12)} = -0.4142136$, $\mathbf{x}^{(12)} = (1, -0.7070918, -0.7071217)^t$

c) $\mu^{(8)} = 5.000029$, $\mathbf{x}^{(8)} = (-0.25000715, 1, -0.25000715)^t$

d) did not converge in 25 iterations

e) $\mu^{(9)} = 1.381959$, $\mathbf{x}^{(9)} = (-0.3819400, -0.2361007, 0.2360191, 1)^t$

f) $\mu^{(6)} = 3.999997$, $\mathbf{x}^{(6)} = (0.9999939, 0.9999999, 0.9999940, 1)^t$

g) $\mu^{(8)} = 3.999996$, $\mathbf{x}^{(8)} = (1, -0.4999959, -0.9999970, -0.5000023)^t$

h) did not converge in 25 iterations

8. The approximate eigenvalues and approximate eigenvectors are:

a) $\mu^{(9)} = 1.000000$, $\mathbf{x}^{(9)} = (-0.1542994, 0.7715207, -0.6172095)^t$

b) $\mu^{(12)} = -0.4142136$, $\mathbf{x}^{(12)} = (-0.7071068, 0.4999894, 0.5000106)^t$

c) $\mu^{(0)} = 2.000000$, $\mathbf{x}^{(0)} = (1, 0, 0)^t$

d) $\mu^{(6)} = 4.961699$, $\mathbf{x}^{(6)} = (-0.4812465, 0.05195336, 0.8750444)^t$

e) $\mu^{(14)} = 2.485863$, $\mathbf{x}^{(14)} = (-0.6096695, 0.6451951, -0.2779286, 0.3671268)^t$

f) $\mu^{(10)} = 3.618034$, $\mathbf{x}^{(10)} = (0.3958550, -0.6404796, 0.6404886, -0.1511924)^t$

g) $\mu^{(7)} = 4.000000$, $\mathbf{x}^{(7)} = (-0.6324604, 0.3162385, 0.6324500, 0.3162183)^t$

h) $\mu^{(6)} = 4.0000000$, $\mathbf{x}^{(6)} = (-0.4999985, -0.5000015, -0.4999985, -0.5000015)^t$

9. a) We have $|\lambda| \leq 6$ for all eigenvalues λ.

b) The approximate eigenvalue is $\lambda_1 = 0.6982681$ with approximate eigenvector $\mathbf{x} = (1, 0.71606, 0.25638, 0.04602)^t$.

c) The eigenvalues are listed in part (d).

d) The characteristic polynomial is $P(\lambda) = \lambda^4 - \frac{1}{4}\lambda - \frac{1}{16}$ and the eigenvalues are $\lambda_1 = 0.6976684972$, $\lambda_2 = -0.237313308$, $\lambda_3 = -0.2301775942 + 0.56965884i$, and $\lambda_4 = -0.2301775942 - 0.56965884i$.

e) The beetle population should approach zero since A is convergent.

EXERCISE SET 9.4 (*Page* 334)

1. Householder's method produces the following tridiagonal matrices.

a) $\begin{bmatrix} 12.00000 & -10.77033 & 0.0 \\ -10.77033 & 3.862069 & 5.344828 \\ 0.0 & 5.344828 & 7.137931 \end{bmatrix}$ **b)** $\begin{bmatrix} 2.0000000 & 1.414214 & 0.0 \\ 1.414214 & 1.000000 & 0.0 \\ 0.0 & 0.0 & 3.0 \end{bmatrix}$

c) $\begin{bmatrix} 2.0000000 & -1.414214 & 0.0 \\ -1.414214 & 3.000000 & 0.0 \\ 0.0 & 0.0 & 1.000000 \end{bmatrix}$ **d)** $\begin{bmatrix} 1.0000000 & -1.414214 & 0.0 \\ -1.414214 & 1.000000 & 0.0 \\ 0.0 & 0.0 & 1.000000 \end{bmatrix}$

e) $\begin{bmatrix} 2.0 & -1.0 & 0.0 \\ -1.0 & -1.0 & -2.0 \\ 0.0 & -2.0 & 3.0 \end{bmatrix}$ **f)** $\begin{bmatrix} 4.750000 & -2.263846 & 0.0 \\ -2.263846 & 4.475610 & -1.219512 \\ 0.0 & -1.219512 & 5.024390 \end{bmatrix}$

2. Householder's method produces the following tridiagonal matrices.

a) $\begin{bmatrix} 4.0000000 & 1.4142136 & 0 & 0 \\ 1.4142136 & 4.0000000 & 1.4142136 & 0 \\ 0 & 1.4142136 & 4.0000000 & 0 \\ 0 & 0 & 0 & 4.0000000 \end{bmatrix}$

b) $\begin{bmatrix} 5.0000000 & 2.5495098 & 0 & 0 \\ 2.5495098 & 6.38461538 & 2.1407569 & 0 \\ 0 & 2.1407569 & 4.2700005 & 0.6912809 \\ 0 & 0 & 0.6912809 & 4.345384 \end{bmatrix}$

c) $\begin{bmatrix} -4 & -3.162278 & 0 & 0 \\ -3.162278 & -3.8 & 1.691153 & 0 \\ 0 & 1.691153 & -2.274476 & -1.780163 \\ 0 & 0 & -1.780163 & -3.475524 \end{bmatrix}$

d) $\begin{bmatrix} 4 & -1.732051 & 0 & 0 \\ -1.732051 & 2.333333 & -0.9428090 & 0 \\ 0 & -0.9428090 & 3.166667 & -0.8660254 \\ 0 & 0 & -0.8660254 & 1.5 \end{bmatrix}$

e) $\begin{bmatrix} 8.0000000 & -2.3048861 & 0 & 0 & 0 \\ -2.3048861 & 5.9294118 & 1.5022590 & 0 & 0 \\ 0 & 1.5022590 & 1.7714975 & -4.8901511 & 0 \\ 0 & 0 & -4.8901511 & -0.4361218 & -1.0898884 \\ 0 & 0 & 0 & -1.0898884 & 4.7352125 \end{bmatrix}$

f)

$$\begin{bmatrix} 2.0000000 & 1.4142136 & 0 & 0 & 0 \\ 1.4142136 & 3.5000000 & 0.8660254 & 0 & 0 \\ 0 & 0.8660254 & 7.8333333 & 4.7140452 & 0 \\ 0 & 0 & 4.7140452 & 6.6666667 & 1.7320508 \\ 0 & 0 & 0 & 1.7320508 & 6.0000000 \end{bmatrix}$$

EXERCISE SET 9.5 (*Page* 340)

1. Two iterations of the QR method produce the following matrices.

a)

$$A^{(3)} = \begin{bmatrix} 0.6939977 & -0.3759745 & 0.0 \\ -0.3759745 & -1.892417 & -0.03039696 \\ 0.0 & -0.03039696 & 3.413585 \end{bmatrix}$$

b)

$$A^{(3)} = \begin{bmatrix} 4.535466 & 1.212648 & 0.0 \\ 1.212648 & 3.533242 & 3.83 \times 10^{-7} \\ 0.0 & 3.83 \times 10^{-7} & -0.06870782 \end{bmatrix}$$

c)

$$A^{(3)} = \begin{bmatrix} 0.6939977 & 0.3759745 & 0.0 \\ 0.3759745 & 1.892417 & 0.03039696 \\ 0.0 & 0.03039696 & 3.413585 \end{bmatrix}$$

d)

$$A^{(3)} = \begin{bmatrix} 4.679567 & -0.2969009 & 0.0 \\ -2.969009 & 3.052484 & -1.207346 \times 10^{-5} \\ 0.0 & -1.207346 \times 10^{-5} & 1.267949 \end{bmatrix}$$

e)

$$A^{(3)} = \begin{bmatrix} 0.3862092 & 0.4423226 & 0.0 & 0.0 \\ 0.4423226 & 1.787694 & -0.3567744 & 0.0 \\ 0.0 & -0.3567744 & 3.080815 & 3.116382 \times 10^{-5} \\ 0.0 & 0.0 & 3.116382 \times 10^{-5} & 4.745201 \end{bmatrix}$$

f)

$$A^{(3)} = \begin{bmatrix} -2.826365 & 1.130297 & 0.0 & 0.0 \\ 1.130297 & -2.429647 & -0.1734156 & 0.0 \\ 0.0 & -0.1734156 & 0.8172086 & 1.863997 \times 10^{-9} \\ 0.0 & 0.0 & 1.863997 \times 10^{-9} & 3.438803 \end{bmatrix}$$

g)

$$A^{(3)} = \begin{bmatrix} 0.2763388 & 0.1454371 & 0.0 & 0.0 \\ 0.1454371 & 0.4543713 & 0.1020836 & 0.0 \\ 0.0 & 0.1020836 & 1.174446 & -4.36 \times 10^{-5} \\ 0.0 & 0.0 & -4.36 \times 10^{-5} & 0.9948441 \end{bmatrix}$$

h)

$$A^{(3)} = \begin{bmatrix} 6.376729 & -1.988497 & 0.0 & 0.0 \\ -1.988497 & -1.652394 & -0.5187146 & 0.0 \\ 0.0 & -0.5187146 & 1.007133 & 1.14 \times 10^{-6} \\ 0.0 & 0.0 & 1.14 \times 10^{-6} & 2.268531 \end{bmatrix}$$

2. The matrices have the following eigenvalues, accurate to within 10^{-5}.

a) $3.9115033, 2.1294613, -2.0409646$

b) $1.2087122, 5.7912878, 3.0000000$

c) $6.0000000, 2.0000000, 4.0000000, 7.4641016, 0.5358984$

d) $4.0274350, 2.0707128, 3.7275564, 5.7839956, 0.8903002$

3. The matrices in Exercise 1 have the following eigenvalues, accurate to within 10^{-5}.

a) 3.414214, 2.000000, 0.58578644

b) $-0.06870782, 5.346462, 2.722246$

c) 3.41424, 2.000000, 0.5857864

d) 1.267949, 4.732051, 3.000000

e) 4.745281, 3.177283, 1.822717, 0.2547188

f) $3.438803, 0.8275517, -1.488068, -3.778287$

g) 0.9948440, 1.189091, 0.5238224, 0.1922421

h) $2.268531, 1.084364, 6.844621, -2.197517$

4. The matrices in Exercise 1 have the following eigenvalues, accurate to within 10^{-5}.

a) $(-0.7071067, 1, -0.7071067)^t$, $(1, 0, -1)^t$, $(0.7071068, 1, 0.7071068)^t$

b) $(0.1741299, -0.5343539, 1)^t$, $(0.4261735, 1, 0.4601443)^t$,
$(1, -0.2777544, -0.3225491)^t$

c) $(0.7071068, 1, 0.7071068)^t$, $(-1, 0, 1)^t$,
$(-0.7071068, 1, -0.7071068)^t$

d) $(0.2679492, 0.7320508, 1)^t$, $(1, -0.7320508, 0.2679492)^t$,
$(1, 1, -1)^t$

e) $(-0.08029447, -0.3007254, 0.7452812, 1)^t$, $(0.4592880, 1, -0.7179949, 0.8727118)^t$,
$(0.8727118, 0.7179949, 1, -0.4592880)^t$ $(1, -0.7452812, -0.3007254, 0.08029447)^t$

f) $(-0.01289861, -0.07015299, 0.4388026, 1)^t$, $(-0.1018060, -0.2878618, 1, -0.4603102)^t$, $(1, 0.5119322, 0.2259932, -0.05035423)^t$ $(-0.5623391, 1, 0.2159474, -0.03185871)^t$

g) $(-0.1520150, -0.3008950, -0.05155956, 1)^t$, $(0.3627966, 1, 0.7459807, 0.3945081)^t$, $(1, 0.09528962, -0.6907921, 0.1450703)^t$, and $(0.8029403, -0.9884448, 1, -0.1237995)^t$

h) $(-0.2172064, -0.1253620, 0.3108802, 1)^t$, $(1, 0.9718785, -0.4885653, 0.4909285)^t$, $(1, -0.9482070, 0.1976567, 0.03688974)^t$, and $(0.1978041, 0.4086314, 1, -0.2166890)^t$

CHAPTER 10 NUMERICAL SOLUTION OF NONLINEAR SYSTEMS OF EQUATIONS

EXERCISE SET 10.2 (Page 348)

1. a) Graphs of the two equations are shown below.

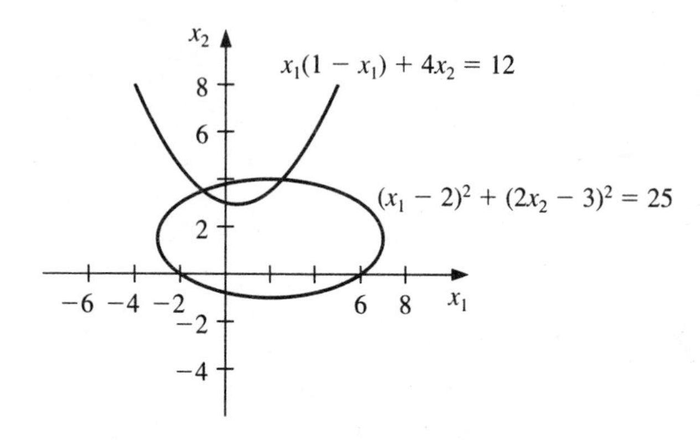

b) With $\mathbf{x}^{(0)}=(0,0)^t$ we have $\mathbf{x}^{(10)}=(-1,3.5)^t$ and with $\mathbf{x}^{(0)}=(6,3)^t$ we have $\mathbf{x}^{(6)}=(2.546947,3.984997)^t$.

2. With $\mathbf{x}^{(0)}=(0.25,0.25)^t$ we have $\mathbf{x}^{(5)}=(0.1212419,0.2711052)^t$.

3. a) With $\mathbf{x}^{(0)}=(0,0)^t$ we have $\mathbf{x}^{(5)}=(0.5,2.0)^t$.

b) With $\mathbf{x}^{(0)}=(1,1)^t$ we have $\mathbf{x}^{(5)}=(0.5,0.8660254)^t$.

c) With $\mathbf{x}^{(0)}=(2,2)^t$ we have $\mathbf{x}^{(6)}=(1.772454,1.772454)^t$.

d) With $\mathbf{x}^{(0)}=(0,0)^t$ we have $\mathbf{x}^{(6)}=(-0.3736982,0.05626649)^t$.

4. a) With $\mathbf{x}^{(0)}=(1,1,1)^t$ we have $\mathbf{x}^{(3)}=(1.036400,1.085707,0.9311914)^t$.

b) With $\mathbf{x}^{(0)}=(0.05,0.2,0.8)^t$ we have $\mathbf{x}^{(3)}=(0,0.1,1)^t$.

c) With $\mathbf{x}^{(0)}=(-1.5,-1.5,0.5)^t$ we have $\mathbf{x}^{(4)}=(-1.456043,-1.664230,0.4224934)^t$.

d) With $\mathbf{x}^{(0)}=(-1.5,-1.5,-5)^t$ we have $\mathbf{x}^{(8)}=(-1.209699,-1.507825,-5.167052)^t$.

5. a) With $\mathbf{x}^{(0)}=(0.1,-0.1,0.1)^t$ we have $\mathbf{x}^{(3)}=(0.4981446,\ -0.1996059,-0.5288259)^t$.

b) With $\mathbf{x}^{(0)}=(1,2,3)^t$ we have $\mathbf{x}^{(16)}=(-0.4021187,\ -0.9302338,8.398353)^t$.
With $\mathbf{x}^{(0)}=(-0.5,-1,2)^t$ we have $\mathbf{x}^{(23)}=(7.869499,\ 0.7214381,-4.980240)^t$.
With $\mathbf{x}^{(0)}=(-0.5,-1,4)^t$ we have $\mathbf{x}^{(14)}=(-0.6398107,\ -1.102016,4.455568)^t$.
With $\mathbf{x}^{(0)}=(-0.5,-1,6)^t$ we have $\mathbf{x}^{(17)}=(-0.9302344,\ -0.4021213,-8.398351)^t$.

c) With $\mathbf{x}^{(0)} = (0.5, 0.5, 0.5, 0.5)^t$ we have $\mathbf{x}^{(14)} = (0.8688753, 0.8688760, 0.8688761, 1.524497)^t$.

d) With $\mathbf{x}^{(0)} = (1,1,1,1)^t$ we have $\mathbf{x}^{(5)} = (0, 0.7071068, 0.7071068, 1)^t$.
With $\mathbf{x}^{(0)} = (1,-1,1,5)^t$ we have $\mathbf{x}^{(6)} = (0.8164966, 0.4082483, -0.4082483, 3)^t$.
With $\mathbf{x}^{(0)} = (1,-1,1,10)^t$ we have $\mathbf{x}^{(5)} = (0.5773503, -0.5773503, 0.5773503, 6)^t$.

6. a) With $\mathbf{x}^{(0)} = (1,1,1)^t$ we have $\mathbf{x}^{(9)} = (0.4999818, 0.01999927, -0.5231013)^t$.

b) With $\mathbf{x}^{(0)} = (2,0,0,0)^t$ we have $\mathbf{x}^{(18)} = (1.000006, 1.000006, 0.9999984, 0.9999984)^t$.

7. Yes, a stable solution occurs when $x_1 = 8000$ and $x_2 = 40000$.

8. a) $k_1 = 8.77125, k_2 = 0.259690, k_3 = -1.37217$ **b)** 15.73 inches

EXERCISE SET 10.3 (*Page* 355)

1. With $\mathbf{x}^{(0)} = (6,3)^t$ we have $\mathbf{x}^{(8)} = (2.546947, 3.984997)^t$.
With $\mathbf{x}^{(0)} = (1,1)^t$ we have $\mathbf{x}^{(13)} = (-1, 3.5)^t$.

2. With $\mathbf{x}^{(0)} = (0.25, 0.25)^t$ we have $\mathbf{x}^{(5)} = (0.1212419, 0.2711052)^t$.

3. a) With $\mathbf{x}^{(0)} = (0,0)^t$ we have $\mathbf{x}^{(7)} = (0.5, 2.0)^t$.

b) With $\mathbf{x}^{(0)} = (1,1)^t$ we have $\mathbf{x}^{(8)} = (0.5, 0.8660254)^t$.

c) With $\mathbf{x}^{(0)} = (2,2)^t$ we have $\mathbf{x}^{(7)} = (1.772454, 1.772454)^t$.

d) With $\mathbf{x}^{(0)} = (0,0)^t$ we have $\mathbf{x}^{(8)} = (-0.3736982, 0.05626649)^t$.

4. a) With $\mathbf{x}^{(0)} = (1,1,1)^t$ we have $\mathbf{x}^{(3)} = (1.036400, 1.085707, 0.9311914)^t$.

b) With $\mathbf{x}^{(0)} = (0.05, 0.2, 0.8)^t$ we have $\mathbf{x}^{(4)} = (0, 0.1, 1)^t$.

c) With $\mathbf{x}^{(0)} = (-1.5, -1.5, 0.5)^t$ we have $\mathbf{x}^{(4)} = (-1.456043, -1.664230, 0.4224934)^t$.

d) With $\mathbf{x}^{(0)} = (-1.5, -1.5, -5)^t$ we have $\mathbf{x}^{(34)} = (-1.640025, -1.762475, -4.347467)^t$.

5. a) With $\mathbf{x}^{(0)} = (0.1, -0.1, 0.1)^t$ we have $\mathbf{x}^{(4)} = (0.4981447, -0.1996059, -0.5288260)^t$.

b) With $\mathbf{x}^{(0)} = (8, 0.75, -5)^t$ we have $\mathbf{x}^{(32)} = (7.869503, 0.7214736, -4.980267)^t$.
With $\mathbf{x}^{(0)} = (-0.6, -1, 4.5)^t$ we have $\mathbf{x}^{(25)} = (-0.6396095, -1.101958, 4.455394)^t$.

c) With $\mathbf{x}^{(0)} = (0.8, 0.8, 0.8, 1.4)^t$ we have $\mathbf{x}^{(17)} = (0.8688769, 0.8688769, 0.8688769, 1.524492)^t$.

d) With $\mathbf{x}^{(0)} = (1,1,1,1)^t$ we have $\mathbf{x}^{(5)} = (0, 0.7071068, 0.7071068, 1)^t$.
With $\mathbf{x}^{(0)} = (1,-1,1,5)^t$ we have $\mathbf{x}^{(8)} = (0.5773498, -0.5773506, 0.5773506, 6)^t$.
With $\mathbf{x}^{(0)} = (1, 0.5, -0.5, 2)^t$ we have $\mathbf{x}^{(6)} = (0.8164966, 0.4082476, -0.4082490, 3)^t$.

6. a) With $\mathbf{x}^{(0)} = (1,1,1)^t$ we have $\mathbf{x}^{(13)} = (0.4999818, 0.01999835, -0.5231013)^t$.

b) The method will likely fail for any choice of $\mathbf{x}^{(0)}$.

EXERCISE SET 10.4 (*Page* 361)

1. **a)** With $\mathbf{x}^{(0)} = (1,1)^t$ we have $\mathbf{x}^{(12)} = (0.498413, 1.98629)^t$.

b) With $\mathbf{x}^{(0)} = (1,1)^t$ we have $\mathbf{x}^{(2)} = (0.501428, 0.869834)^t$.

c) With $\mathbf{x}^{(0)} = (2,2)^t$ we have $\mathbf{x}^{(1)} = (1.73540, 1.80392)^t$.

d) With $\mathbf{x}^{(0)} = (0,0)^t$ we have $\mathbf{x}^{(3)} = (-0.377890, 0.0508119)^t$.

2. Newton's method improves the approximations in Exercise 1 to:

a) $\mathbf{x}^{(3)} = (0.5, 2)^t$

b) $\mathbf{x}^{(3)} = (0.5, 0.8660254)^t$

c) $\mathbf{x}^{(3)} = (1.772454, 1.772454)^t$

d) $\mathbf{x}^{(3)} = (-0.3736982, 0.05626649)^t$

3. **a)** With $\mathbf{x}^{(0)} = (0,0,0)^t$ we have $\mathbf{x}^{(18)} = (1.03865, 1.07904, 0.928979)^t$.

b) With $\mathbf{x}^{(0)} = (0,0,0)^t$ we have $\mathbf{x}^{(2)} = (-0.0200570, 0.0901966, 0.994681)^t$.

c) With $\mathbf{x}^{(0)} = (0,0,0)^t$ we have $\mathbf{x}^{(17)} = (-1.60120, -1.20804, 0.752611)^t$.

d) With $\mathbf{x}^{(0)} = (1,1,1)^t$ we have $\mathbf{x}^{(3)} = (-0.002049049, 0.0075156, 0.15814)^t$.

e) With $\mathbf{x}^{(0)} = (-0.5,-1,2)^t$ we have $\mathbf{x}^{(23)} = (-1.03044, -1.17457, 1.68028)^t$.

f) With $\mathbf{x}^{(0)} = (1,1,1,1)^t$ we have $\mathbf{x}^{(8)} = (0.00953301, 0.700762, 0.706167, 1.09239)^t$.
With $\mathbf{x}^{(0)} = (1,-1,1,5)^t$ we have $\mathbf{x}^{(4)} = (0.587303, -0.587303, 0.587303, 6.01399)^t$.
With $\mathbf{x}^{(0)} = (1,-1,1,2)^t$ we have $\mathbf{x}^{(4)} = (0,0,0,2.49517)^t$.

4. Newton's method improves the approximations in Exercise 3 to:

a) $\mathbf{x}^{(2)} = (1.036400, 1.085707, 0.9311914)^t$

b) $\mathbf{x}^{(3)} = (0, 0.1, 1)^t$

c) $\mathbf{x}^{(5)} = (-1.456043, -1.664230, 0.4224934)^t$

d) $\mathbf{x}^{(20)} = (0, 0, -1.000007)^t$

e) $\mathbf{x}^{(16)} = (1.758835, -11.58335, -5.397068)^t$

f) $\mathbf{x}^{(3)} = (0, 0.7071068, 0.7071068, 1)^t$
$\mathbf{x}^{(3)} = (0.5773503, -0.5773503, 0.5773503, 6)^t$, singular Jacobian

5. **a)** With $\mathbf{x}^{(0)} = (0.1, 0.1)^t$ we have $\mathbf{x}^{(8)} = (5.343082, -0.6262875)^t$ and $g(\mathbf{x}^{(8)}) = 0.006995494$.

b) With $\mathbf{x}^{(0)} = (0,0)^t$ we have $\mathbf{x}^{(13)} = (0.6157412, 0.3768953)^t$ and $g(\mathbf{x}^{(13)}) = 0.1481574$.

c) With $\mathbf{x}^{(0)} = (0,0,0)^t$ we have $\mathbf{x}^{(5)} = (-0.6633785, 0.3145720, 0.5000740)^t$ and $g(\mathbf{x}^{(5)}) = 0.6921548$.

d) With $\mathbf{x}^{(0)} = (1,1,1)^t$ we have $\mathbf{x}^{(4)} = (0.04022273, 0.01592477, 0.01594401)^t$ and $g(\mathbf{x}^{(4)}) = 1.010003$.

CHAPTER 11 BOUNDARY-VALUE PROBLEMS FOR ORDINARY DIFFERENTIAL EQUATIONS

EXERCISE SET 11.2 (*Page 368*)

1. The Linear Shooting method gives the results in the following tables.

a)

i	x_i	w_{1i}
1	0.333333	0.5311664
2	0.666667	1.153515

b)

i	x_i	w_{1i}
1	0.25	0.3937095
2	0.50	0.8240948
3	0.75	1.337160

2. The Linear Shooting method gives the results in the following tables.

a)

i	x_i	w_{1i}	w_{2i}
1	0.78539816	−0.28245222	0.14141733

b)

i	x_i	w_{1i}	w_{2i}
1	0.52359878	−0.30974274	0.06363321
2	1.04719755	−0.23655220	0.20972171

3. The Linear Shooting method gives the results in the following tables.

a)

i	x_i	w_{1i}
3	0.3	0.7833204
6	0.6	0.6023521
9	0.9	0.8568906

b)

i	x_i	w_{1i}
5	1.0	0.00865076
10	2.0	0.00007484
15	3.0	0.00000065

c)

i	x_i	w_{1i}
5	1.25	0.1676179
10	1.50	0.4581901
15	1.75	0.6077718
20	2.00	0.6931460

d)

i	x_i	w_{1i}
3	0.3	−0.5185754
6	0.6	−0.2195271
9	0.9	−0.0406577

e)

i	x_i	w_{1i}
3	1.3	0.0655336
6	1.6	0.0774590
9	1.9	0.0305619

f)

i	x_i	w_{1i}
3	1.515485	10.094751
6	2.030969	17.547048
9	2.546454	17.380532

4. The Linear Shooting method gives the results in the following tables.

a)

i	x_i	w_{1i}	w_{2i}
1	0.15707963	1.05248506	0.25267869
2	0.31415927	1.07905470	0.08492370
3	0.47123890	1.07905469	−0.08492234
4	0.62831853	1.05248505	−0.25267729

b)

i	x_i	w_{1i}	w_{2i}
1	0.15707963	−0.06061198	−0.29443007
2	0.31415927	−0.09117479	−0.09251254
3	0.47123890	−0.08959214	0.11091096
4	0.62831853	−0.05748564	0.29239128

c)

i	x_i	w_{1i}	w_{2i}
5	1.25000000	0.64314227	0.28800448
10	1.50000000	0.68324209	0.07407700
15	1.75000000	0.69226853	0.01166358

d)

i	x_i	w_{1i}	w_{2i}
3	0.60000000	−0.71219638	−1.82098025
5	1.00000000	−1.64068454	−2.81187530
8	1.60000000	−3.52051591	−2.83551329

5. The Linear Shooting method with $h = 0.05$ gives the following results.

i	x_i	w_{1i}
6	0.3	0.04990547
10	0.5	0.00676467
16	0.8	0.00033755

The Linear Shooting method with $h = 0.1$ gives the following results.

i	x_i	w_{1i}
3	0.3	0.05273437
5	0.5	0.00741571
8	0.8	0.00038976

6. For Eq. (11.1), let $u_1(x) = y$ and $u_2(x) = y'$. Then

$$u_1'(x) = u_2(x), \quad a \le x \le b, \quad u_1(a) = \alpha$$

and

$$u_2'(x) = p(x)u_2(x) + q(x)u_1(x) + r(x), \quad a \le x \le b, \quad u_2(a) = 0.$$

For Eq. (11.2), let $v_1(x) = y$ and $v_2(x) = y'$. Then

$$v_1'(x) = v_2(x), \quad a \le x \le b, \quad v_1(a) = 0$$

and

$$v_2'(x) = p(x)v_2(x) + q(x)v_1(x), \quad a \le x \le b, \quad v_2(a) = 1.$$

7. a)

x	$w(x)$
24	0.0071265
48	0.011427
60	0.011999
72	0.011427
96	0.0071265
120	0.0

b) Yes

c) The actual solution is within code, but the numerical solution does not indicate this.

8. a) The approximate potential with $N = 20$ is $u(3) \approx 36.66669$.

b) The approximate potential with $N = 40$ is $u(3) \approx 36.66666$.

c) The actual potential is $u(3) = 36.\overline{6}$.

EXERCISE SET 11.3 (*Page* 373)

1. The Linear Finite-Difference method gives the results in the following tables.

a)

i	x	w_i
1	0.333333	0.5343259
2	0.666667	1.1579818

b)

i	x	w_i
1	0.25	0.3951247
2	0.50	0.8265306
3	0.75	1.3395692

2. The Linear Finite-Difference method gives the results in the following tables.

a)

i	x_i	w_{1i}
1	0.78539816	−0.15366578

b)

i	x_i	w_{1i}
1	0.52359878	−0.31018048
2	1.04719755	−0.23634609

3. The Linear Finite-Difference method gives the results in the following tables.

a)

i	x_i	w_i
2	0.2	1.018096
5	0.5	0.5942743
7	0.7	0.6514520

b)

i	x_i	w_i
5	1.0	6.332971×10^{-3}
10	2.0	4.010654×10^{-5}
15	3.0	2.539917×10^{-7}

c)

i	x_i	w_i
5	1.25	0.16797186
10	1.50	0.45842388
15	1.75	0.60787334

d)

i	x_i	w_i
3	0.3	−0.5183084
6	0.6	−0.2192657
9	0.9	−0.0405748

e)

i	x_i	w_i
3	1.3	0.0654387
6	1.6	0.0773936
9	1.9	0.0305465

f)

i	x_i	w_i
3	1.515485	1.904530
6	2.030969	5.415273
9	2.546454	11.935402

4. The Linear Finite-Difference method gives the results in the following tables.

a)

i	x_i	w_i
1	0.15707963	1.05260081
2	0.31415927	1.07922974
3	0.47123890	1.07922974
4	0.62831853	1.05260081

b)

i	x_i	w_i
1	0.15707963	−0.06141845
2	0.31415927	−0.09240491
3	0.47123890	−0.09080499
4	0.62831853	−0.05825827

c)

i	x_i	w_i
5	1.25	0.64328225
10	1.50	0.68332838
15	1.75	0.69230217

d)

i	x_i	w_i
3	0.6	−0.70664241
5	1.0	−1.63674050
8	1.6	−3.52936107

5. The Linear Finite-Difference method gives the results in the following tables.

i	x_i	$w_i(h = 0.1)$
3	0.3	0.05572807
6	0.6	0.00310518
9	0.9	0.00016516

i	x_i	$w_i(h = 0.05)$
6	0.3	0.05132396
12	0.6	0.00263406
18	0.9	0.00013340

6. The Linear Finite-Difference method with extrapolation gives the results in the following tables.

a)

x_i	$w_i(h = 0.05)$	$w_i(h = 0.025)$	EXT_{1i}	EXT_{2i}	EXT_{3i}
0.2	1.021161	1.021937	1.022181	1.022195	1.022196
0.5	0.5964505	0.5970101	0.5971746	0.5971965	0.5971980
0.7	0.6525620	0.6528496	0.6529312	0.6529451	0.6529460

b)

x_i	$w_i(h = 0.05)$	$w_i(h = 0.025)$	EXT_{1i}	EXT_{2i}	EXT_{3i}
1.0	7.579267×10^{-3}	7.894285×10^{-3}	7.994696×10^{-3}	7.999289×10^{-3}	7.999591×10^{-3}
2.0	5.744566×10^{-5}	6.231994×10^{-5}	6.322536×10^{-5}	6.394466×10^{-5}	6.399261×10^{-5}
3.0	4.353907×10^{-7}	4.919569×10^{-7}	4.958567×10^{-7}	5.108119×10^{-7}	5.118088×10^{-7}
4.0	3.289033×10^{-9}	3.870230×10^{-9}	3.850686×10^{-9}	4.063959×10^{-9}	4.078177×10^{-9}

c)

x_i	$w_i(h = 0.025)$	$w_i(h = 0.0125)$	EXT_{1i}	EXT_{2i}	EXT_{3i}
1.25	0.1677325	0.1677398	0.1676508	0.1677423	0.1677483
1.50	0.4582796	0.4583252	0.4582315	0.4583404	0.4583477
1.75	0.6078227	0.6078653	0.6078058	0.6078795	0.6078844

7.

i	x_i	w_i
10	10.0	0.1098549
20	20.0	0.1761424
25	25.0	0.1849608
30	30.0	0.1761424
40	40.0	0.1098549

EXERCISE SET 11.4 (*Page* 379)

1. The Nonlinear Shooting method gives $w_1 = 0.405991 \approx \ln 1.5 = 0.405465$

2.

i	x_i	w_{1i}	w_{2i}
0	1.00000000	0.25000000	
1	1.25000000	0.23529995	−0.05534017
2	1.50000000	0.22223398	−0.04935887
3	1.75000000	0.21054420	−0.04429643
4	2.00000000	0.20002431	−0.03997377

3. The Nonlinear Shooting method gives the results in the following tables. To apply the method in parts d) and f), we need to redefine the initial value of T_k to be 1.5 and 2, respectively.

a)

i	x_i	w_{1i}
3	1.3	0.434783
6	1.6	0.384615
9	1.9	0.344828

b)

i	x_i	w_{1i}
3	1.3	2.069249
6	1.6	2.225015
9	1.9	2.426322

c)

i	x_i	w_{1i}
3	1.3	1.031597
6	1.6	1.095005
9	1.9	1.168170

d)

i	x_i	w_{1i}
5	0.3926991	0.6600925
10	0.7853982	1.1319596
15	1.1780973	1.4454742
20	1.5707963	1.5707414

e)

i	x_i	w_{1i}
5	1.25	−0.7272908
10	1.50	−0.8000371
15	1.75	−0.8889473

f)

i	x_i	w_{1i}
5	1.25	0.4358290
10	1.50	1.3684496
15	1.75	2.9992010
20	2.00	5.5451958

4. The method gives the results in the following tables. To apply the method in parts b), d), and f), we need to define the initial approximations for t_k to be -0.5 and 0.5, 1.5 and 2.5, and 0.5 and 1.81832, respectively.

a)

i	x_i	w_{1i}
3	1.3	0.4347720
6	1.6	0.3845947
9	1.9	0.3447969

b)

i	x_i	w_{1i}
3	1.3	2.0692491
6	1.6	2.2250137
9	1.9	2.4263174

c)

i	x_i	w_{1i}
3	1.3	1.0315965
6	1.6	1.0950047
9	1.9	1.1681698

d)

i	x_i	w_{1i}
5	0.3926991	0.6600904
10	0.7853982	1.1319554
15	1.1780973	1.4454656

e)

i	x_i	w_{1i}
5	1.25	−0.7272728
10	1.50	−0.8000001
15	1.75	−0.8888891

f)

i	x_i	w_{1i}
5	1.25	0.4358261
10	1.50	1.3684417
15	1.75	2.9991849
20	2.00	5.5451671

EXERCISE SET 11.5 (*Page* 381)

1. The Nonlinear Finite-Difference method gives the following results.

i	x_i	w_i
1	1.5	0.406760

2. The Nonlinear Finite-Difference method gives the following results.

i	x_i	w_i
1	1.25000000	0.23530108
2	1.50000000	0.22223064
3	1.75000000	0.21053210

3. The Nonlinear Finite-Difference method gives the results in the following tables.

a)

i	x_i	w_i
3	1.3	0.434796
6	1.6	0.384627
9	1.9	0.344831

b)

i	x_i	w_i
3	1.3	2.0694081
6	1.6	2.2250937
9	1.9	2.4263387

c)

i	x_i	w_i
3	1.3	1.031970
6	1.6	1.095321
9	1.9	1.168271

d)

i	x_i	w_i
3	0.471239	0.766923
6	0.942478	1.275944
9	1.413717	1.548057

e)

i	x_i	w_i
5	1.25	−0.727281
10	1.50	−0.800013
15	1.75	−0.888900

f)

i	x_i	w_i
5	1.25	0.434598
10	1.50	1.366212
15	1.75	2.996934

4. **a)**

x_i	$w_i(h = 0.1)$	$w_i(h = 0.05)$	$w_i(h = 0.025)$	Extrapolated
1.2	0.4545563	0.4545481	0.4545460	0.4545453
1.5	0.4000130	0.4000032	0.4000008	0.4000000
1.7	0.3703797	0.3703727	0.3703709	0.3703703

b)

x_i	$w_i(h = 0.1)$	$w_i(h = 0.05)$	$w_i(h = 0.025)$	Extrapolated
1.1	2.00923190	2.00912670	2.00909989	2.00909091
1.2	2.03351579	2.03337957	2.03334493	2.03333333
1.3	2.06940807	2.06927563	2.06924202	2.06923077
1.4	2.11443858	2.11432434	2.11429540	2.11428572
1.5	2.16678956	2.16669768	2.16667444	2.16666667
1.6	2.22509370	2.22502362	2.22500592	2.22500000
1.7	2.28830287	2.28825232	2.28823956	2.28823530
1.8	2.35560005	2.35556676	2.35555836	2.35555555
1.9	2.42633872	2.42632157	2.42631724	2.42631579

EXERCISE SET 11.6 (*Page* 392)

1. The Piecewise Linear method gives $\phi(x) = -0.07713274\phi_1(x) - 0.07442678\phi_2(x)$.

2. The Piecewise Linear method gives $\phi(x) = -0.2552629\phi_1(x) - 0.1633565\phi_2(x)$.

3. The Piecewise Linear method gives the results in the following tables.

a)

i	x_i	$\phi(x_i)$
3	0.3	4.0908778×10^{-2}
6	0.6	5.8304182×10^{-2}
9	0.9	2.6538926×10^{-2}

b)

i	x_i	$\phi(x_i)$
3	0.3	−0.212333
6	0.6	−0.241333
9	0.9	−0.090333

c)

i	x_i	$\phi(x_i)$
3	0.3	0.1815153
6	0.6	0.1805512
9	0.9	0.05936480

d)

i	x_i	$\phi(x_i)$
5	0.25	−0.3586155
10	0.50	−0.5348645
15	0.75	−0.4510389

e)

i	x_i	$\phi(x_i)$
5	0.25	−0.1846134
10	0.50	−0.2737099
15	0.75	−0.2285169
20	1.00	0.0000000

f)

i	x_i	$\phi(x_i)$
3	0.3	-5.738551×10^{-2}
6	0.6	-4.974304×10^{-2}
9	0.9	-1.478349×10^{-2}

4. The Cubic Spline method gives the results in the following tables.

a)

i	c_i
0	-4.38127×10^{-3}
1	-5.29927×10^{-2}
2	-6.86142×10^{-2}
3	-5.02453×10^{-2}
4	-3.10118×10^{-3}

b)

i	c_i
0	-1.38889×10^{-2}
1	−0.138889
2	−0.180556
3	−0.138889
4	-1.38889×10^{-2}

5. The Cubic Spline method gives the results in the following tables.

a)

i	x_i	$\phi(x_i)$
3	0.3	4.0878126×10^{-2}
6	0.6	5.8259848×10^{-2}
9	0.9	2.6518200×10^{-2}

b)

i	x_i	$\phi(x_i)$
3	0.3	−0.2100000
6	0.6	−0.2400000
9	0.9	−0.0900000

c)

i	x_i	$\phi(x_i)$
3	0.3	0.1814269
6	0.6	0.1804753
9	0.9	0.05934321

d)

i	x_i	$\phi(x_i)$
5	0.25	−0.3585641
10	0.50	−0.5347803
15	0.75	−0.4509614

e)

i	x_i	$\phi(x_i)$
5	0.25	−0.1845203
10	0.50	−0.2735857
15	0.75	−0.2284204

f)

i	x_i	$\phi(x_i)$
3	0.3	-5.754895×10^{-2}
6	0.6	-4.985645×10^{-2}
9	0.9	-1.481176×10^{-2}

6. The transformation gives $F(x) = f(x) + (\beta - \alpha)p'(x) - (\beta x + \alpha(1-x))q(x)$

7. The Piecewise Linear method gives the results in the following table.

i	x_i	$\phi(x_i)$
3	0.3	−0.06959856
6	0.6	−0.07197026
9	0.9	−0.02454043

8. The Cubic Spline method gives the results in the following table.

x_i	$\phi_i(x)$
0.3	1.040810
0.5	1.106531
0.8	1.249338

9. A change in variable $t = (x - a)/(b - a)$ gives the boundary value problem

$$-\frac{d}{dt}(p((b-a)t+a)y') + (b-a)^2 q((b-a)t+a)y = (b-a)^2 f((b-a)t+a),$$

where $0 < t < 1$, $y(0) = \alpha$, and $y(1) = \beta$. Then Exercise 6 can be used.

Chapter 12 Numerical Methods for Partial Differential Equations

EXERCISE SET 12.2 (*Page* 403)

1. The Poisson Equation Finite-Difference method gives the following results.

i	j	x_i	y_j	w_{ij}
1	1	0.5	0.5	0.0
1	2	0.5	1.0	0.25
1	3	0.5	1.5	1.0

2. The Poisson Equation Finite-Difference method gives the following results.

i	j	x_i	y_j	w_{ij}
1	1	1.16666667	0.33333333	0.38610055
1	2	1.16666667	0.66666667	0.59095111
2	1	1.33333333	0.33333333	0.63513150
2	2	1.33333333	0.66666667	0.79842236
3	1	1.50000000	0.33333333	0.85833078
3	2	1.50000000	0.66666667	0.99098460
4	1	1.66666667	0.33333333	1.06025052
4	2	1.66666667	0.66666667	1.16984761
5	1	1.83333333	0.33333333	1.24443628
5	2	1.83333333	0.66666667	1.33631576

3. The Poisson Equation Finite-Difference method gives the following results.

a)

i	j	x_i	y_j	w_{ij}
2	2	0.4	0.4	0.159999
2	4	0.4	0.8	0.319999
4	2	0.8	0.4	0.320000
4	4	0.8	0.8	0.640000

b)

i	j	x_i	y_j	w_{ij}
2	2	0.4	0.4	1.8579248
2	4	0.4	0.8	1.2399321
4	2	0.8	0.4	6.1038748
4	4	0.8	0.8	3.8335165

c)

i	j	x_i	y_j	w_{ij}
4	3	0.8	0.3	1.27136
4	7	0.8	0.7	1.75084
8	3	1.6	0.3	1.61675
8	7	1.6	0.7	3.06587

d)

i	j	x_i	y_j	w_{ij}
2	1	1.256637	0.3141593	0.2951912
2	3	1.256637	0.9424778	0.1830968
4	1	2.513274	0.3141593	−0.7721915
4	3	2.513274	0.9424778	−0.4785097

e)

i	j	x_i	y_j	w_{ij}
2	2	1.2	1.2	0.5251533
4	4	1.4	1.4	1.3190830
6	6	1.6	1.6	2.4065150
8	8	1.8	1.8	3.8088995

f)

i	j	x_i	y_j	w_{ij}
2	2	0.2	0.2	−0.9171063
4	4	0.4	0.4	−0.9558396
6	6	0.6	0.6	−0.9672948
8	8	0.8	0.8	−0.8841996

4. The approximate potential at some typical points is shown in the following table.

i	j	x_i	y_j	w_{ij}
1	4	0.1	0.4	88
2	1	0.2	0.1	66
4	2	0.4	0.2	88

5. Approximations for the temperature are given in the following table.

i	j	x_i	y_j	w_{ij}
5	9	2.0	3.0	5.957716
8	3	3.2	1.0	7.915441
10	9	4.0	3.0	4.678240
12	12	4.8	4.0	2.059610

EXERCISE SET 12.3 (*Page* 414)

1. The Heat Equation Backward-Difference method gives the following results.

a)

i	j	x_i	t_j	w_{ij}
1	1	0.5	0.05	0.632952
2	1	1.0	0.05	0.895129
3	1	1.5	0.05	0.632952
1	2	0.5	0.1	0.566574
2	2	1.0	0.1	0.801256
3	2	1.5	0.1	0.566574

b)

i	j	x_i	t_j	w_{ij}
1	1	1/3	0.05	1.59728
2	1	2/3	0.05	−1.59728
1	2	1/3	0.1	1.47300
2	2	2/3	0.1	−1.47300

2. The Crank-Nicolson method gives the following results.

a)

i	j	x_i	t_j	w_{ij}
1	1	0.5	0.05	0.628848
2	1	1.0	0.05	0.889326
3	1	1.5	0.05	0.628848
1	2	0.5	0.1	0.559251
2	2	1.0	0.1	0.790901
3	2	1.5	0.1	0.559252

b)

i	j	x_i	t_j	w_{ij}
1	1	1/3	0.05	1.591825
2	1	2/3	0.05	−1.591825
1	2	1/3	0.1	1.462951
2	2	2/3	0.1	−1.462951

3. The Forward-Difference method gives the following results.

a) For $h = 0.1$ and $k = 0.01$:

i	j	x_i	t_j	w_{ij}
4	50	0.4	0.5	-9.3352×10^8
10	50	1.0	0.5	-9.1860×10^8
17	50	1.7	0.5	2.6047×10^8

For $h = 0.1$ and $k = 0.005$:

i	j	x_i	t_j	w_{ij}
4	100	0.4	0.5	$3.6726805 \times 10^{-10}$
10	100	1.0	0.5	$1.0503891 \times 10^{-17}$
17	100	1.7	0.5	$-5.942522 \times 10^{-10}$

b) For $h = \frac{\pi}{10}$ and $k = 0.05$:

i	j	x_i	t_j	w_{ij}
3	10	0.9424778	0.5	0.4921015
6	10	1.8849556	0.5	0.5785001
9	10	2.8274334	0.5	0.1879661

c)

i	j	x_i	t_j	w_{ij}
4	10	0.8	0.4	1.166142
8	10	1.6	0.4	1.252404
12	10	2.4	0.4	0.4681804
16	10	3.2	0.4	−0.1027628

d)

i	j	x_i	t_j	w_{ij}
2	10	0.2	0.4	0.3921147
4	10	0.4	0.4	0.6344550
6	10	0.6	0.4	0.6344550
8	10	0.8	0.4	0.3921148

4. a) For $h = 0.1$ and $k = 0.01$:

i	j	x_i	t_j	w_{ij}
4	50	0.4	0.5	5.5507362×10^{-8}
10	50	1.0	0.5	$9.1889966 \times 10^{-17}$
17	50	1.7	0.5	-8.981280×10^{-8}

For $h = 0.1$ and $k = 0.005$:

i	j	x_i	t_j	w_{ij}
4	100	0.4	0.5	1.5089062×10^{-8}
10	100	1.0	0.5	$3.0735816 \times 10^{-17}$
17	100	1.7	0.5	-2.441461×10^{-8}

b) For $h = \pi/10$ and $k = 0.05$:

i	j	x_i	t_j	w_{ij}
3	10	0.94247780	0.5	0.49331321
6	10	1.88495559	0.5	0.57992446
9	10	2.82743339	0.5	0.18842888

c)

i	j	x_i	t_j	w_{ij}
4	10	0.8	0.4	1.1767516
8	10	1.6	0.4	1.2594950
12	10	2.4	0.4	0.46281339
16	10	3.2	0.4	−0.1123064

d)

i	j	x_i	t_j	w_{ij}
2	10	0.2	0.4	0.39834082
4	10	0.4	0.4	0.64452899
6	10	0.6	0.4	0.64452899
8	10	0.8	0.4	0.39834082

5. The Crank-Nicolson method gives the following results.

a) For $h = 0.1$ and $k = 0.01$:

i	j	x_i	t_j	w_{ij}
4	50	0.4	0.5	2.3541×10^{-9}
10	50	1.0	0.5	1.7610×10^{-17}
17	50	1.7	0.5	-3.8090×10^{-9}

For $h = 0.1$ and $k = 0.01$:

i	j	x_i	y_j	w_{ij}
4	100	0.4	0.5	2.8156746×10^{-9}
10	100	1.0	0.5	$2.4953437 \times 10^{-17}$
17	100	1.7	0.5	-4.555857×10^{-9}

b) For $h = \pi/10$ and $k = 0.05$:

i	j	x_i	t_j	w_{ij}
2	10	0.628319	0.5	0.357938
5	10	1.570796	0.5	0.608960
8	10	2.513274	0.5	0.357938

c)

i	j	x_i	t_j	w_{ij}
5	10	1	0.4	1.312434
10	10	2	0.4	0.9050248
15	10	3	0.4	−0.03253811

d)

i	j	x_i	t_j	w_{ij}
3	10	0.3	0.4	0.5440574
5	10	0.5	0.4	0.6724913
7	10	0.7	0.4	0.5440568

6. We need to change the method to incorporate the following steps.

STEP Set

$$t = jk;$$
$$z_1 = (w_1 + kF(h))/l_1.$$

STEP For $i = 2, \ldots, m-1$ set

$$z_i = (w_i + kF(ih) + \lambda z_{i-1})/l_i.$$

7. For the modified Backward-Difference method we have the results in the following table.

i	j	x_i	t_j	w_{ij}
3	25	0.3	0.25	0.2883455
5	25	0.5	0.25	0.3468410
8	25	0.8	0.25	0.2169213

8. We need to modify the method to incorporate the following steps.

STEP Set

$t = jk$;
$w_0 = \phi(t)$;
$z_1 = (w_1 + \lambda w_0)/l_1$.
$w_m = \psi(t)$.

STEP For $i = 2, \ldots, m-2$ set

$z_i = (w_i + \lambda z_{i-1})/l_i$;
$z_{m-1} = (w_{m-1} + \lambda w_m + \lambda z_{m-2})/l_{m-1}$.

STEP OUTPUT (t);
For $i = 0, \ldots, m$ set $x = ih$;
OUTPUT (x, w_i).

9. For the modified Backward-Difference method we have the results in the following table.

i	j	x_i	t_j	w_{ij}
3	10	0.3	0.225	1.207730
6	10	0.75	0.225	1.836564
10	10	1.35	0.225	0.6928342

10. **a)** The approximate temperature at some typical points is given below.

i	j	r_i	t_j	w_{ij}
1	20	0.6	10	137.6753
2	20	0.7	10	245.9678
3	20	0.8	10	340.2862
4	20	0.9	10	424.1537

b) The strain is approximately $I = 1242.537$.

11.

i	j	x_i	t_j	w_{ij}
2	10	200	5	1.478828×10^7
5	10	500	5	4.334451×10^6
8	10	800	5	1.478828×10^7

EXERCISE SET 12.4 (*Page* 421)

1. The Wave Equation Finite-Difference method gives the following results.

i	j	x_i	t_j	w_{ij}
2	4	0.25	1.0	−0.7071068
3	4	0.50	1.0	−1.0000000
4	4	0.75	1.0	−0.7071068

2. The Wave Equation Finite-Difference method gives the following results.

i	j	x_i	t_j	$w(x_i)$
2	4	0.125	0.5	0.48428862
3	4	0.250	0.5	0.00000000
4	4	0.375	0.5	−0.48428862

3. The Wave Equation Finite-Difference method with $h = \frac{\pi}{10}$ and $k = 0.05$ gives the following results.

i	j	x_i	t_j	w_{ij}
2	10	$\frac{\pi}{5}$	0.5	0.516393
5	10	$\frac{\pi}{2}$	0.5	0.878541
8	10	$\frac{4\pi}{5}$	0.5	0.516393

The Wave Equation Finite-Difference method with $h = \frac{\pi}{20}$ and $k = 0.1$ gives the following results.

i	j	x_i	t_j	w_{ij}
4	5	$\frac{\pi}{5}$	0.5	0.515916
10	5	$\frac{\pi}{2}$	0.5	0.877729
16	5	$\frac{4\pi}{5}$	0.5	0.515916

The Wave Equation Finite-Difference method with $h = \frac{\pi}{20}$ and $k = 0.05$ gives the following results.

i	j	x_i	t_j	w_{ij}
4	10	$\frac{\pi}{5}$	0.5	0.515960
10	10	$\frac{\pi}{2}$	0.5	0.877804
16	10	$\frac{4\pi}{5}$	0.5	0.515960

4. The Wave Equation Finite-Difference method gives the following results.

i	j	x_i	t_j	w_{ij}
2	3	0.2	0.3	0.6729902
5	3	0.5	0.3	6.317389×10^{-16}
7	3	0.7	0.3	−0.6729902

5. The Wave Equation Finite-Difference method gives the following results.

i	j	x_i	t_j	w_{ij}
2	5	0.2	0.5	−1
5	5	0.5	0.5	0
7	5	0.7	0.5	1

6. a) The air pressure for the open pipe is $p(0.5, 0.5) \approx 0.9$ and $p(0.5, 1.0) \approx 2.7$.
   b) The air pressure for the closed pipe is $p(0.5, 0.5) \approx 0.9$ and $p(0.5, 1.0) \approx 0.9187927$.

7. Approximate voltages and currents are given in the following table.

i	j	x_i	t_j	Voltage	Current
5	2	50	0.2	77.782	3.88909
12	2	120	0.2	104.62	−1.69959
18	2	180	0.2	33.992	−5.23081
5	5	50	0.5	77.782	3.88909
12	5	120	0.5	104.62	−1.69959
18	5	180	0.5	33.992	−5.23081

EXERCISE SET 12.5 (*Page* 432)

1. With $E_1 = (0.25, 0.75)$, $E_2 = (0, 1)$, $E_3 = (0.5, 0.5)$, and $E_4 = (0, 0.5)$, the basis functions are

$$\phi_1(x,y) = \begin{cases} 4x & \text{on } T_1 \\ -2+4y & \text{on } T_2 \end{cases}, \quad \phi_2(x,y) = \begin{cases} -1-2x+2y & \text{on } T_1 \\ 0 & \text{on } T_2 \end{cases},$$

$$\phi_3(x,y) = \begin{cases} 0 & \text{on } T_1 \\ 1+2x-2y & \text{on } T_2 \end{cases}, \quad \phi_4(x,y) = \begin{cases} 2-2x-2y & \text{on } T_1 \\ 2-2x-2y & \text{on } T_2 \end{cases}$$

and $\gamma_1 = 0.323825, \gamma_2 = 0, \gamma_3 = 1.0000$, and $\gamma_4 = 0$.

2. With $E_1 = (0.25, 0.75), E_2 = (0, 1), E_3 = (0.5, 0.5), E_4 = (0, 0.5), E_5 = (0, 0.75)$, and $E_6 = (0.25, 0.5)$ the following results are obtained:

i	j	$a_j^{(i)}$	$b_j^{(i)}$	$c_j^{(i)}$
1	1	0	4	0
1	2	−3	0	4
1	3	4	−4	−4
2	1	−2	0	4
2	2	−1	4	0
2	3	4	−4	−4
3	1	0	4	0
3	2	3	0	−4
3	3	−2	4	4
4	1	−2	0	4
4	2	1	−4	0
4	3	2	4	−4

So $\gamma_1 = 0.3461969$, $\gamma_2 = 0$, $\gamma_3 = 1.0$, $\gamma_4 = 0, \gamma_5 = 0$, and $\gamma_6 = 0.5$.

3. The Finite-Element method with $K = 8, N = 8, M = 32, n = 9, m = 25$, and $NL = 0$ gives the following results:

$\gamma_1 = 0.511023$, $\gamma_2 = 0.720476$, $\gamma_3 = 0.507898$, $\gamma_4 = 0.720475$, $\gamma_5 = 1.01885$,
$\gamma_6 = 0.720476$, $\gamma_7 = 0.507897$, $\gamma_8 = 0.720476$, $\gamma_9 = 0.511023$,
and
$\gamma_i = 0$, for each $i = 10, 11, \ldots, 25$.

$u(0.125, 0.125) \approx 0.614187$, $u(0.125, 0.25) \approx 0.690343$, $u(0.25, 0.125) \approx 0.690343$, and $u(0.25, 0.25) \approx 0.720475$. (See the diagram.)

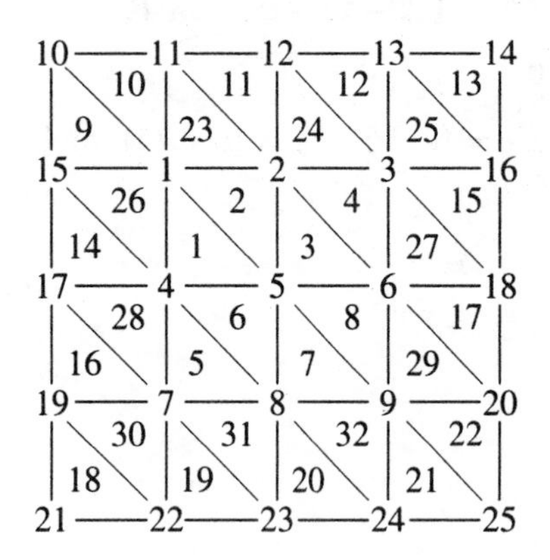

4. The Finite-Element method with $K = 8, N = 22, M = 32, n = 25, m = 25$, and $NL = 16$ gives the following results:

$\gamma_1 = -0.489695$, $\gamma_2 = 0.016323$, $\gamma_3 = 0.524240$, $\gamma_4 = 0.016325$, $\gamma_5 = 0.008685$,
$\gamma_6 = 0.016324$, $\gamma_7 = 0.524242$, $\gamma_8 = 0.016325$, $\gamma_9 = -0.489695$, $\gamma_{10} = -1.06913$,
$\gamma_{11} = -0.684308$, $\gamma_{12} = 0.058157$, $\gamma_{13} = 0.752868$, $\gamma_{14} = 0.962799$, $\gamma_{15} = -0.684308$,
$\gamma_{16} = 0.752869$, $\gamma_{17} = 0.058157$, $\gamma_{18} = 0.058157$, $\gamma_{19} = 0.752868$, $\gamma_{20} = -0.684310$,
$\gamma_{21} = 0.962801$, $\gamma_{22} = 0.752870$, $\gamma_{23} = 0.058157$, $\gamma_{24} = -0.684312$,
and $\gamma_{25} = -1.06913$.

$u(0.125, 0.125) \approx 0.270284$, $u(0.125, 0.25) \approx -0.238595$, $u(0.25, 0.125) \approx -0.238595$, and $u(0.25, 0.25) \approx 0.016323$

5. The Finite-Element method with $K = 0, N = 12, M = 32, n = 20, m = 27$, and $NL = 14$ gives the following results:

$\gamma_1 = 21.40335$, $\gamma_2 = 19.87372$, $\gamma_3 = 19.10019$, $\gamma_4 = 18.85895$, $\gamma_5 = 19.08533$,
$\gamma_6 = 19.84115$, $\gamma_7 = 21.34694$, $\gamma_8 = 24.19855$, $\gamma_9 = 24.16799$, $\gamma_{10} = 27.55237$,
$\gamma_{11} = 25.11508$, $\gamma_{12} = 22.92824$, $\gamma_{13} = 21.39741$, $\gamma_{14} = 20.52179$, $\gamma_{15} = 20.23334$,
$\gamma_{16} = 20.50056$, $\gamma_{17} = 21.35070$, $\gamma_{18} = 22.84663$, $\gamma_{19} = 24.98178$, $\gamma_{20} = 27.41907$,
$\gamma_{21} = 15$, $\gamma_{22} = 15$, $\gamma_{23} = 15$, $\gamma_{24} = 15$, $\gamma_{25} = 15$,
$\gamma_{26} = 15$, and $\gamma_{27} = 15$.

$u(1,0) \approx 22.92824$, $u(4,0) \approx 22.84663$, and $u\left(\frac{5}{2}, \frac{\sqrt{3}}{2}\right) \approx 18.85895$. (See the diagram.)

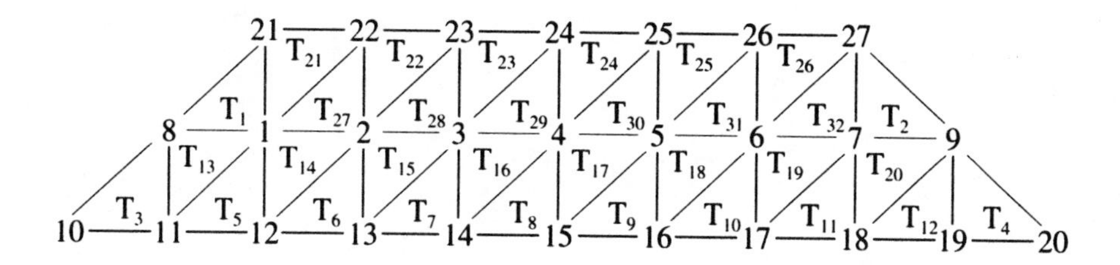